Professor Richard Wiseman started his working life as an award-winning professional magician. After completing an initial degree in psychology, he spent four years testing psychics and mediums as part of his doctorate at the Koestler Parapsychology Unit, Edinburgh University. For the past twenty years he has investigated the psychology of the paranormal, spending sleepless nights in haunted castles, investigating gurus in India, attempting to talk with the dead, and examining psychic dogs. Professor Wiseman has published over fifty academic papers on the paranormal and is a fellow of the Committee for the Scientific Investigation of Claims of the Paranormal. He has spoken at The Royal Society, Microsoft, Caltech, Google, and The Royal Institution. Over a million people have taken part in Professor Wiseman's mass participation experiments, and his YouTube channel has received over 10 million views. A survey conducted by *Times Higher Education* revealed that he was the psychologist most frequently quoted in the British media. Professor Wiseman is also listed as a 'must follow' scientist by The Celebrity Twitter Directory, and in 2010 the *Independent on Sunday* named him as one of the top 100 people who make Britain a better place to live.

Also by Richard Wiseman

THE LUCK FACTOR

DID YOU SPOT THE GORILLA?

QUIRKOLOGY

59 SECONDS

Professor Richard Wiseman

PARANORMALITY

Why We See What Isn't There

MACMILLAN

First published 2011 by Macmillan

an imprint of Pan Macmillan, a division of Macmillan Publishers Limited
Pan Macmillan, 20 New Wharf Road, London N1 9RR
Basingstoke and Oxford
Associated companies throughout the world
www.panmacmillan.com

ISBN 978-0-230-75298-6

3 5 7 9 8 6 4

A CIP catalogue record for this book is available from
the British Library.

Printed in the UK by CPI Mackays, Chatham ME5 8TD

Visit www.panmacmillan.com to read more about all our books
and to buy them. You will also find features, author interviews and
news of any author events, and you can sign up for e-newsletters
so that you're always first to hear about our new releases.

To Jeff

Interactive Tags

Several sections of the book contain these designs:

www.richardwiseman.com/paranormality/Welcome.html

They are called QR tags and they allow you to experience exclusive short films and audio clips on your smartphone. Simply open any barcode scanning application, point your camera at the design, and your phone will automatically link to the additional content. If you don't have a smartphone, the web address for this additional material is printed below each tag.

Contents

6. MIND CONTROL

In which we climb inside the head of the world's greatest thought-reader, discover whether hypnotists can make us act against our will, infiltrate some cults, learn how to avoid being brainwashed and investigate the psychology of persuasion.

7. PROPHECY

In which we find out whether Abraham Lincoln really did foresee his own death, learn how to control our dreams and delve deep into the remarkable world of sleep science.

CONCLUSION

In which we find out why we are all wired for the weird and contemplate the nature of wonder.

THE INSTANT SUPERHERO KIT

A parting gift: Six psychological techniques to impress your friends and family.

A Quick Test Before We Begin

This book contains lots of tests, experiments, exercises and demonstrations. Here is the first of them. Take a quick look at the inkblot below.

What does the image look like to you?

Many thanks. As we will discover later, the thoughts that have just crossed your mind reveal a great deal about you.

Introduction

In which we learn what happened when a supposedly psychic dog was put to the test, and begin our journey into a world where everything appears possible and nothing is quite as it seems.

As I gazed deep into the eyes of Jaytee, several thoughts passed through my mind. Was this cute little terrier really psychic? If not, how had he managed to make headlines around the world? And if he could predict the future, did he already know if our experiment would be a success? At that precise moment, Jaytee gave a small cough, leaned forward and vomited on my shoes.

My quality time with Jaytee took place about a decade ago. I was in my early thirties and conducting an experiment to discover whether this supposedly psychic terrier really could predict when his owners would return home. By then I had already spent over ten years investigating a variety of alleged paranormal phenomena, spending sleepless nights in supposedly haunted houses, testing mediums and psychics, and carrying out laboratory experiments into telepathy.

This fascination with the impossible started when I was eight years old and I saw my first magic trick. My grandfather had me mark my initials on a coin, made the coin disappear, and then revealed that it had been magically transported into a sealed box. A few weeks later he explained the secret to the supposed miracle and I was hooked. For the next few years I found out everything I could about the dark arts of magic and deception. I searched second-hand bookshops for obscure works on sleight of hand, joined a local magic club, and performed for friends and family. By my teens I had a couple of hundred shows under my belt and had become one of the youngest members of the prestigious Magic Circle.

In order to successfully deceive an audience, magicians have to understand how you think and behave. More specifically, they need to know how to make you misperceive what is happening inches from your nose, prevent you from thinking about certain solutions to tricks, and persuade you to misremember what has happened right in front of your eyes. After fooling people on a twice-nightly basis for several years I became fascinated with these aspects of human behaviour, and eventually decided to enrol for a psychology degree at University College London.

Like most magicians, I was deeply sceptical about the existence of paranormal phenomena, and had confined them to a mental file-drawer labelled 'not true, but fun to talk about at parties'. Then, when I was just coming to the end of the first year of my psychology degree, a chance event changed everything. One day I happened to turn on the television in my student digs and caught the end of a programme about science and the supernatural. A young psychologist named Sue Blackmore popped up on the screen and explained that she was also fascinated by things that allegedly go bump in the night. Then she said something that had a huge impact on my career. Instead of examining whether such phenomena were genuine, she explained that she thought it more worthwhile to investigate why people experienced these strange sensations. Why did mothers think that they were in telepathic communication with their children? Why did people believe that they had seen a ghost? Why were some people so certain that their destiny was written in the stars? Suddenly, the penny dropped. Before then I hadn't seriously considered carrying out any research into paranormal phenomena. After all, why would I waste my time looking at the possible reality of things that probably didn't exist? However, Sue's comments made me realize that such work could be worthwhile if I were to move away from the existence of the phenomena themselves and instead focus on the deep and fascinating psychology that lay behind people's beliefs and experiences.

As I delved deeper I discovered that Sue was not the only researcher to have adopted this approach to the paranormal. In fact, throughout history a handful of researchers have dedicated their lives to discovering what supposedly paranormal phenomena tell us about our behaviour, beliefs and brain. Daring to take a walk on the weird side, these pioneering mavericks have carried out some of the strangest research ever conducted, including removing the head of the world's top thought-reader, infiltrating several cults, attempting to weigh the souls of the dying, and testing a talking mongoose. Just as the mysterious Wizard of Oz turned out to be a man behind a curtain pushing buttons and pulling levers, so their work has yielded surprising and important insights into the psychology of everyday life and the human psyche.

My investigation into the allegedly psychic terrier Jaytee is a good example of the approach.

Before becoming the highly successful self-help guru that he is today, Paul McKenna hosted a television series about the paranormal. I was invited to be one of the resident scientists on the show, offering my opinion on a whole range of remarkable performances, experiments and events. It was a mixed bag. One week a man appeared to generate sparks from his fingertips, while another time Paul invited millions of viewers to psychically influence the national lottery by concentrating on seven specific numbers during the draw (three of the numbers came up).

One episode involved an especially interesting film about a terrier called Jaytee. According to the film, Jaytee had the uncanny ability to predict when his owner, Pam, was returning home. Pam lived with her parents and they had noticed Jaytee seemed to reliably signal their daughter's homecoming by sitting in the window. A national newspaper had published an article on Jaytee's amazing ability and an Austrian television company had conducted an initial experiment with him. The test was shown on Paul McKenna's programme and involved one film crew following Pam as she walked around her local town centre while a second crew continuously filmed Jaytee in her parents' house. When Pam decided to return home Jaytee went to the window and remained there until his owner arrived. Pam, Jaytee and I were all on the show and chatted about the film. I said that I thought it was very curious, and Pam kindly invited me to conduct a more formal examination of her apparently psychic dog.

A few months later my research assistant, Matthew Smith, and I found ourselves driving to Ramsbottom in north-west England to test Jaytee. We all met and everything seemed to be going well. Pam was very friendly, Matthew and I liked Jaytee, and Jaytee seemed to like us.

During the first test Matthew and Pam drove to a public house about 8 miles away and, once there, used a random number generator to select a time to head back – 9 p.m. Meanwhile, I continuously filmed Jaytee's favourite window so that we would have a complete record of his behaviour there. When Pam and Mat returned from the bar we rewound the film and eagerly observed Jaytee's behaviour. Interestingly, the terrier was at the window at the allotted time. So far, so good. However, when we looked at the remainder of the film, Jaytee's apparent skills started to unravel. It turned out that he was something of a fan of the window, visiting it 13 times during the experiment. During a second trial the following day, Jaytee visited the window 12 times. It seemed his time in the window was not the clear-cut signal that the clip from Austrian television suggested. Pam explained that summer was perhaps the wrong time for the experiment because of the many distractions, including the local bitch being on heat and the coming of the fishmonger.

In December we returned to Ramsbottom and conducted another two trials. In the first session Jaytee made four separate trips to the window and one of them was about ten minutes before Matthew and Pam set off home. Close, but no cigar. On the final trial Jaytee made eight trips to the window. One of them was just as Matthew and Pam started their return trip, but he only spent a few seconds there before running into the garden and vomiting on my shoes.

All in all, not exactly overwhelming evidence for animal magic.[1] However, the interesting question is not whether animals really have psychic gifts, but rather, why might

Field footage of Jaytee test
www.richardwiseman.com/paranormality/Jaytee.html

people come to believe that they have a psychic bond with their pet? The answer tells us a great deal about one of the most fundamental ways in which we think about the world.

In 1967, psychologist husband and wife team Loren and Jean Chapman, from the University of Wisconsin, conducted a now classic experiment.[2] The study involved a form of psychiatric assessment that was popular in the 1960s called the 'Draw A Person Test'. According to clinicians at the time, it was possible to detect all sorts of possible problems, such as paranoia, repressed sexuality and depression from an individual's drawing of a typical person. The Chapmans, however, were not so sure that the test stood up to scrutiny. After all, many of the alleged relationships, such as paranoid people making drawings with large eyes, seemed to fit surprisingly well with the stereotypes that the public carry around in their heads, and so the Chapmans wondered whether the alleged patterns were actually in the minds of the clinicians. To test their idea, a group of students was presented with drawings of people made by psychiatric patients, along with a brief description of their symptoms, such as 'He is suspicious', 'He is worried about not being manly enough', 'He is worried about sexual impotence'. After looking through the pairings of pictures and words, the volunteers were asked whether they had noticed any patterns in the data. Interestingly, the volunteers reported the same types of patterns that professionals had been using for years. They thought, for example, that paranoid people draw atypical eyes, those with issues surrounding their manliness produced broad-shouldered figures and that small sexual organs were indicative of impotence-related matters.

There was just one small problem. The Chapmans had randomly paired up the drawing and symptoms, so there were no real patterns in the data. The volunteers had seen the invisible. The Chapmans' work completely discredited the 'Draw A Person Test' and, more importantly, revealed an important insight into the human psyche. Our beliefs do not sit passively in our brains waiting to be confirmed or contradicted by incoming information. Instead, they play a key role in shaping how we see the world. This is especially true when faced with coincidences. We are remarkably good at paying attention to events that coincide, especially when they sup-port our beliefs. In the Chapmans' experiment, volunteers already believed that paranoid people would produce drawings with large eyes, and so noticed instances when a particular person's drawing actually had large eyes and played down the images from paranoid individuals that had perfectly normal eyes.

The same principle applies to matters of the paranormal. We all like to think that we have untapped psychic potential and get excited when we think of a friend, the telephone rings, and they're on the other end of the line. In doing so, we are forgetting all the occasions when we thought about that friend, the telephone rang, and it was a double-glazing salesman. Or all the times you weren't thinking about the friend and they unexpectedly telephoned. Similarly, if we have a dream that reflects the following day's events, we are quick to claim the gift of prophecy, but in doing so we are ignoring all of the times when our dreams didn't come true. It is the same with animal magic. If we believe that owners have a psychic bond with their pets, we pay attention to when an animal seems to predict their owner's homecoming, and forget when the animal made a prediction but was wrong, or failed to foresee a return.

Perhaps more importantly, the same mechanism also leads us astray with matters of health. In the mid-1990s researchers Donald Redelmeier and Amos Tversky decided to investigate the possible link between arthritic pain and the weather.[3] For thousands of years people have convinced themselves that their arthritis flares up with certain changes in temperature, barometric pressure, and humidity. To find out if this was really the case, Redelmeier and Tversky had a group of rheumatoid arthritis sufferers rate their pain levels twice a month for over a year. The research team then obtained detailed information about the local temperature, barometric pressure and humidity over the same time period. All of the patients were convinced that there was a relationship between the weather and their pain. However, the data showed that their condition was completely unrelated to the weather patterns. Once again, they had focused on the times when high levels of pain were associated with especially odd weather patterns, forgotten about the times when this was not the case, and erroneously concluded that the two were related.

Similarly, we might hear about someone who was miraculously healed after praying, forget about those who were healed without prayer or prayed but were not healed, and incorrectly conclude that prayer works. Or we might read about someone who was cured of cancer after eating lots of oranges, forget those who were cured without oranges or consumed oranges but weren't cured, and end up believing that oranges help cure cancer.

The effect can even play a role in promoting racism, with

people seeing images of those from ethnic minorities engaged in acts of violence, forgetting about the individuals from minorities who are law-abiding citizens and the violent people from non-minority backgrounds, and concluding that those from minorities are especially likely to commit crime.

My research into Jaytee started off with an investigation into a supposedly psychic dog and ended up revealing a great deal about one of the most fundamental ways in which we misperceive the world. This illustrates why I find supernatural science so fascinating. Each journey takes you on a voyage into the unknown where you have no idea who you are going to meet or what you are going to find.

We are about to embark on an expedition deep into this hitherto hidden world of supernatural science. In a series of fantastical tales, we will meet a colourful cast of characters, go backstage with expert illusionists, observe charismatic cult leaders in action, and attend mindboggling séances. Each adventure will reveal unique and surprising insights into the hidden psychology behind your everyday life, including, for example, how you have evolved to be afraid of things that go bump in the night, how your unconscious is far more power-ful than previously imagined, and how your mind can be controlled by others. The journey is going to be far more than a passive sightseeing trip. Along the way you will be urged to roll up your sleeves and take part in several experiments. Each of these tests offers an opportunity to explore the more mys-terious side of your psyche, encouraging you, for example, to measure your powers of intuition, assess how suggestible you are, and discover if you are a natural-born liar.

It is almost time to depart. Prepare to enter a world where anything appears possible and yet nothing is ever quite what it seems. A world where the truth really is stranger than fiction. A world that I have had the pleasure of calling home for the past twenty years.

Hurry now, there's a storm brewing, and we are about to begin our journey into a world far more wonderful than Oz . . .

1. FORTUNE-TELLING

In which we meet the mysterious 'Mr D', visit the
non-existent town of Lake Wobegon, find out how to
convince strangers that we know all about them
and discover who we really are.

For reasons that will soon become apparent, it wouldn't be fair to give Mr D's real name. Born in the north of England in 1934, this remarkable man spent much of his life working as a professional psychic and developed a considerable reputation for highly accurate readings. When I was studying at Edinburgh University, Mr D contacted me and asked whether I would be interested in watching him give some readings. I immediately accepted the kind offer and invited Mr D to the University so that I could film him at work. A few weeks later the two of us met in the foyer of the Psychology Department. I showed him into my laboratory and explained that I had lined up several volunteers who were eager to take part in a psychic reading. Mr D quietly set up his table, took out his Tarot cards and crystal ball, and waited for his first guinea pig. A few moments later the door opened and in walked a 43-year-old barmaid named Lisa. I pressed the 'record' button on the video camera and retreated to the other side of a two-way mirror.

Mr D knew nothing about Lisa before the reading. He started by asking her to hold out her right hand, palm up. After carefully examining her palm with a horn-handled magnifying glass, Mr D started to describe her personality. Within seconds Lisa was nodding and smiling. He next asked her to shuffle a deck of Tarot cards and then place them in the centre of the table. Mr D turned over one card after another and spoke about each in turn. Within a few minutes he told Lisa that she had a brother and described his career in considerable detail. A few moments later Mr D said that he thought Lisa had recently broken up from a long-term relationship.

Lisa's reading lasted around ten minutes. When she left the laboratory I interviewed her about what she thought about her time with Mr D. Lisa was extremely impressed, and explained how Mr D had been correct about her personality, recent relationship difficulties and brother's career. When I asked Lisa to rate the accuracy of Mr D's reading she gave it top marks.

Throughout the morning several other people came away equally convinced that Mr D possessed uncanny powers. After a spot of lunch, Mr D watched the recordings of his readings and explained more about his abilities. It proved a fascinating and eye-opening experience. In just a few hours Mr D not only provided a rare glimpse into the world of the professional psychic, but also revealed how almost anyone could learn to develop such powers. At the end of the day Mr D packed away his Tarot cards and said goodbye. Unfortunately, I never met Mr D again because he suffered a sudden and fatal heart attack a few years later. However, the day that

Laboratory footage of Mr D at work
www.richardwiseman.com/paranormality/MrD.html

I spent with him lives on in my mind, and we will return to the secret behind his seemingly magical gift of insight later in the chapter.

Every year millions of people visit psychics and come away completely convinced that these individuals have the ability to see deep within their souls. Are they kidding themselves, the victims of elaborate scams, or is something genuinely spooky going on? To find out, a small number of researchers have put the alleged paranormal powers of psychics and mediums under the microscope, of whom the most notable investigator is magician and arch-sceptic, James Randi.

Séance on a Warm
Wednesday Afternoon

Randall James Hamilton Zwinge was born in Toronto in 1928.[1] When he was 12 years old, he happened to catch a matinée performance by a well-known American magician named Harry Blackstone Sr. The bug bit deep, and Zwinge found out as much as he could about the secretive world of magic and eventually started to perform on a regular basis.

Like many magicians, Zwinge was a tad sceptical about matters paranormal. When he was 15 he went along to his local spiritualist church and was disgusted by what he witnessed. People in the congregation were encouraged to bring along sealed envelopes containing questions to their deceased loved ones. The ministers then secretly read the messages and created a fake reply from the 'dead'. Zwinge attempted to expose the deception, but upset the ministers and ended up spending time at the local police station.

Unperturbed, he eventually grew a goatee, legally changed his name to James 'The Amazing' Randi, and embarked on a long and colourful career as a professional magician and escapologist. Over the years Randi has been involved in a series of headline-grabbing projects, including remaining in a sealed metal coffin for 104 minutes (breaking Houdini's record by just over ten minutes), clocking up 22 appearances on Johnny Carson's *The Tonight Show*, featuring in an episode of *Happy Days*, escaping from a straitjacket while hanging upside-down over Niagara Falls, and appearing to behead rock legend Alice Cooper on a nightly basis.

In tandem with his magic career, Randi continued his crusade against paranormal chicanery. His investigations gained such momentum and notoriety that in 1996 he established the James Randi Educational Foundation. The website promotes itself as 'an educational resource on the paranormal, pseudoscientific and the supernatural' and it also offers a bold challenge to would-be psychics or those professing to have paranormal powers. A million dollar challenge to be precise.

In the late 1960s Randi appeared on a radio chat show explaining why he thought that those claiming paranormal powers were either deluding themselves or deceiving others. One panellist, a parapsychologist, suggested that he put his money where his mouth was by offering a cash prize to anyone who could show that they had genuine psychic abilities. Randi took up the challenge and put up $1,000. Over the years Randi's offer grew to $100,000 and then, in the late 1990s, a wealthy supporter of his Foundation increased the prize fund to one million dollars to anyone who can demonstrate the existence of paranormal abilities to the satisfaction of an independent panel (so far, no one has). But for over a decade this opportunity to become an instant millionaire has attracted a steady stream of applicants, including psychics who claim to be able to guess the order of shuffled decks of cards, dowsers who say they can use bent coat hangers and forked sticks to discover underground water, and even a woman who tried to use the power of her mind to make strangers urinate. That, too, was a failure . . .

In 2008 a British medium called Patricia Putt applied for Randi's million-dollar challenge. Putt was convinced that she was able to garner information about the living by chatting with their deceased friends and relatives. Randi asked me and Chris French, a Professor of Psychology at Goldsmiths College in London, to test Putt's abilities.[2]

Putt lives in Essex and is an experienced medium who has given both personal and group readings for several years. According to her website, much of this work has been carried out with the invaluable assistance of her Egyptian spirit guide 'Ankhara', whom she first encountered while participating in a regression hypnotherapy session. Putt's website also describes many instances where she has apparently provided undeniable proof of the spirit world, as well as listing several television and radio programmes that have enlisted her services.

After much discussion, Putt, French and I agreed on the details of the test. It was to take place on one day and involve ten volunteers. Putt would not know any of these people in advance, and would attempt to contact a deceased friend or relative of each volunteer, and then use this spirit to determine information about the volunteer's personality and life.

The big day arrived. Each of the volunteers was scheduled to arrive at French's laboratory at different times throughout the day. To minimize the possibility of Putt picking up any information about the volunteers by the way they looked or dressed, French had them remove any watches and jewellery, don a full-length black cape, and put on a black balaclava.

Each volunteer was shown into the test room and asked to sit in a chair facing a wall. Putt then came in, sat at a desk on the opposite side of the room and attempted to make contact with the spirit world. Once she thought she had a direct line to the dead, Putt located a spirit that knew the person and then quietly wrote down information about the volunteer. My role in the test was to bring Putt in and out of the test room at appropriate times, stay with her as she attempted to contact the spirits, and to generally keep her company throughout the day. Putt and I spent much of the time between the sessions chatting. At one point I asked her if there was a downside to working as a professional psychic. Without a hint of irony she explained how annoying it was when people made an appointment to see her but then failed to show up.

A volunteer takes part in Patricia Putt's test.

After Putt had completed all ten sessions the volunteers were asked to return to the test room. They were each given transcriptions of all of the readings that Putt had made that day and were asked to look through them and identify the reading that seemed to apply to them. If Putt really had the powers she claimed, the volunteers should have had an easy time. For example, let's imagine that one of them had been brought up in the country, had spent a significant amount of time travelling in France, and had recently married an actor. If Putt really did have a direct line to the spirit world, then she might have mentioned a childhood surrounded by greenery, the strong whiff of brie, or the phrase, 'darling, it was a triumph'. Once the person saw those comments they would instantly know that that reading was intended for them, and so would have no problem choosing it from the pack. In order for Putt to pass the test, five or more of the volunteers had to correctly identify their reading.

Each volunteer carefully examined Putt's readings and identified the one that they found most accurate. We all then gathered in French's office to see how Putt had scored. Volunteer One had chosen a reading that had been meant for Volunteer Seven. The reading selected by Volunteer Two was actually made when Volunteer Six was sitting in front of Putt. And so it went on. In fact, none of the volunteers correctly identified their reading. Putt was stunned by the result but has vowed to return with a new and improved claim.[3]

Interview with Prof. Chris French
www.richardwiseman.com/paranormality/ChrisFrench.html

You could argue that Putt failed because she agreed to work under an artificial set of conditions. After all, unless she gets a gig at an introverted amateur Batman look-a-like convention, she will rarely be asked to produce readings for people who are dressed in a black cape, wearing a black balaclava, and facing away from her. The problem is that other experiments conducted in more natural settings have yielded the same result.

In the early 1980s psychologists Hendrik Boerenkamp and Sybo Schouten from the University of Utrecht spent five years studying the alleged paranormal powers of 12 well-respected Dutch psychics.[4]

The researchers visited each psychic in their home several times each year ('Is he expecting you?'), showed the psychic a photograph of someone that they had never met and asked them to provide information about that person. They also carried out exactly the same procedure with a group of randomly selected people who didn't claim to be psychic. After recording and analysing over 10,000 statements, the researchers concluded that the allegedly paranormal powers of the psychics failed to outperform the random guesses made by the non-psychic control group, and that neither group produced impressive hit rates.

These types of failed studies are not the exception, they are the norm.[5]

For over a century researchers have tested the claims of mediums and psychics and found them wanting. Indeed, after reviewing this vast amount of work, Sybo Schouten concluded that the psychics' performance was simply no better than chance. It seems that when it comes to psychics and mediums, Randi's million-dollar prize is safe.

The conundrum is that surveys suggest that around one in six people believe they have received an accurate reading from an alleged psychic.[6]

To solve the mystery it is necessary to learn the secrets of the psychic readers. There are several ways of doing this. You could, for example, spend several weeks on a psychic development programme attempting to open your inner eye. Or you could enrol on a month-long course at a college for mediumship and try to tune into the dead. Alternatively, you could save yourself a great deal of time and effort by forgetting all about that. Whether intentionally or unintentionally, most mediums and psychics use a fascinating set of psychological techniques to give the impression that they have a magical insight into the past, present and future. These techniques are referred to as 'cold reading', and they reveal an important insight into the fundamental nature of our everyday interactions. To find out about them we are going to spend some more time with a familiar friend of ours.

Revealing the Mysterious Mr D

Before continuing our journey into the psychology of psychic readings I would like you to take the following two-part psychological test.

First, imagine that the illustration below represents an aerial view of a large sandpit. Next, imagine that someone has randomly chosen a place in the pit and buried some treasure there. You have just one opportunity to dig down and find the treasure. Without thinking about it too much, place an 'X' in the sandpit to indicate where you would dig.

Second, simply think of one geometric shape inside another. Many thanks. We will return to your answers later on.

At the start of the chapter I described how Mr D once visited Edinburgh University and demonstrated his amazing abilities. In reading after reading, complete strangers sat down opposite him and left convinced that he knew all about them. One of the most impressive readings was given to Lisa, who had no idea how Mr D had come up with accurate information about her personality, her brother's career and her recent relationship difficulties.

As you might have guessed by now, Mr D did not possess genuine paranormal powers. In fact, he had spent much of his life using cold reading to fake psychic ability, and was happy to reveal the tricks of his trade. Mr D used six psychological techniques to appear to achieve the impossible.[7]

To understand the first of these we need to travel to the non-existent town of Lake Wobegon.

1. Flattery Will Get You Everywhere

In the mid-1980s American writer and humorist Garrison Keillor created a fictional town called Lake Wobegon. According to Keillor, Lake Wobegon is located in the centre of Minnesota but can't be found on maps because of the incompetence of nineteenth-century surveyors. When describing the townsfolk, Keillor noted that 'all the women are strong, all the men are good-looking, and all the children are above average'. Although written in jest, Keillor's comment reflects a key psychological principle now referred to as the 'Lake Wobegon effect'.

Much of the time you make rational decisions. However, under certain circumstances your brain trips you up, and you suddenly let go of logic. Psychologists discovered that a major cause of irrationality revolves around a curious phenomenon known as the 'egocentric bias'. Nearly all of us have fragile egos and use various techniques to protect ourselves from the harsh reality of the outside world. We are highly skilled at convincing ourselves that we are responsible for the success in our lives, but equally good at blaming failures on other people. We fool ourselves into believing that we are unique, possess above average abilities and skills, and are likely to experience more than our fair share of good fortune in the future. The effects of egocentric thinking can be dramatic. In perhaps the best-known example, researchers asked each member of long-term couples to estimate the percentage of the housework they carried out. The combined total from almost every pair exceeded 100 per cent. Each had displayed an egocentric bias by focusing on their own work and down-playing their partner's contribution.

For the most part, this egotism is good for you. It makes you feel positive about yourself, motivates you to get up in the morning, helps you deal with the slings and arrows of outrageous fortune, and persuades you to carry on when the going gets tough. For example, research has shown that people are unrealistically optimistic about both their personality and abilities. 94 per cent of people think that they have an above average sense of humour, 80 per cent of drivers say that they are more skilled than the average driver (remarkably, this is even true of those that are in hospital because they have been involved in a road accident), and 75 per cent of business people see themselves as more ethical than the average businessman.[8] It is the same when it comes to personality. Present people with any positive trait and they are quick to tick the 'yes, that's me' box, leading to the vast majority of people irrationally believing themselves to be far more cooperative, considerate, responsible, friendly, reliable, resourceful, polite and dependable than the average person. These delusions are the price that we pay for the happiness, success and resilience that we enjoy in the rest of our lives.

A good cold reader exploits your egocentric thinking by telling you how wonderful you are. Mr D's readings were full of flattery. After just a few moments glancing at Lisa's palm, Mr D told her that she had a good imagination, possessed lots of creative flair and had an eye for detail. A few moments later Lisa learned that she could have been a psychic because she was very intuitive, had the unusual ability of giving her opinions about people without hurting their feelings, and was a very caring person. Each time she heard compliments like these, the Lake Wobegon effect kicked in, leaving Lisa with no explanation for Mr D's allegedly accurate insights into her personality.

But cold readings are not just about visiting Lake Wobegon. They also involve the little-known 'Dartmouth Indians versus the Princeton Tigers' effect.

2. Seeing What You Want To See

In 1951 American University football team the Dartmouth Indians played the Princeton Tigers. It was an especially rough game, with Princeton's quarterback suffering a broken nose and a Dartmouth player being stretchered off with a broken leg. However, newspapers from each of the two Universities presented very different descriptions of the game, with the Dartmouth journalists describing how the Princeton players had caused the problems, while the Princeton journalists were convinced that the Dartmouth team were to blame. Was this simply media bias? Intrigued, social psychologists Albert Hastorf and Hadley Cantril tracked down Dartmouth and Princeton students who had been at the game and interviewed them about what they had seen.[9] Even though they had been watching exactly the same event, the two groups focused on different aspects of the action, resulting in vastly differing views about what had happened. For example, when asked whether the Dartmouth team started the rough play, 36 per cent of the Dartmouth students ticked the 'yes' box versus 86 per cent of the Princeton students. Likewise, just 8 per cent of the Dartmouth students thought that the Dartmouth team were unnecessarily rough, compared to 35 per cent of the Princeton students. Researchers have discovered that the same phenomenon (referred to as 'selective memory') occurs in many different contexts – when people with strong beliefs are presented with ambiguous information relevant to their views, they will see what they want to see.

This 'Dartmouth Indians versus the Princeton Tigers' effect also helps explain the success of Lisa's reading. When Mr D first looked at her hand, he spoke about many aspects of Lisa's personality, with lots of his statements predicting both one trait and the exact opposite. Lisa was told that she was both highly sensitive yet also very down-to-earth, and that although many people saw her as shy in reality she wasn't afraid to speak her mind. In the same way that the Dartmouth and Princeton students remembered the parts of the football game that matched their preconceptions, so Lisa focused on the aspects of Mr D's statements that she believed applied to her and paid almost no attention to all of the incorrect information. Lisa heard what she wanted to hear, and came away convinced of Mr D's mysterious powers.

Following hot on the heels of the 'Lake Wobegon' effect, and 'Dartmouth Indians versus the Princeton Tigers' effect, is the third key principle of cold reading, the 'Doctor Fox'effect.

3. The Creation of Meaning

Look at the symbol below.

13

If the letter 'A' is placed on one side of the symbol, and the letter 'C' on the other, most people have no problem interpreting the symbol as a 'B'.

ABC

However, if the number '12' is placed above the symbol, and the number '14' below, the mysterious symbol shape-shifts into a '13'.

12
13
14

Or you could be especially sneaky, and place the letters 'A' and 'C' to the left and right, and the numbers '12' and '14' above and below, and suddenly the symbol continually flips between being the letter 'B' and the number '13'.

All of this nicely illustrates a fundamental quirk of the human perceptual system. Given the right context, people are skilled at instantly and unconsciously seeing meaning in a meaningless shape. The same principle helps people see all sorts of images in inkblots, clouds and toasted waffles. Stare at these random shapes for long enough and suddenly objects, faces and figures will start to emerge.

The same process occurs during our everyday conversations. When you chat with someone, the two of you try your best to convey your thoughts to one another. Some of your comments might be somewhat vague and ambiguous, but the human brain is pretty good at inferring meaning from the context of the conversation, and so all is well. However, this vital process can go into overdrive, causing you to hear meaning where there is none.

In the 1970s Donald Naftulin and his colleagues from the University of Southern California demonstrated the power of this principle in dramatic fashion.[10] Naftulin wrote a completely meaningless lecture on the relationship between mathematics and human behaviour, arranged for an actor to present the talk at an education conference, and then asked the audience of psychiatrists, psychologists, and social workers what they thought. Prior to the talk Naftulin had the actor carefully rehearse his lines and coached him on how to deal with the thirty-minute question and answer session by using 'double talk, neologisms, non sequiturs, and contradictory statements'. At the conference Naftulin introduced the actor as 'Dr Myron L. Fox' and briefly reviewed his impressive, but entirely fictitious, curriculum vitae. For the next hour and a half the audience were bombarded with meaningless drivel and contradictory statements. At the end of the session Naftulin handed out a questionnaire and asked everyone for feedback.

In the same way that you saw a meaningless symbol a few moments ago but interpreted it as either the letter 'B' or number '13', so the audience perceived great wisdom in Dr Fox's nonsense. According to those in attendance, Dr Fox had given an 'excellent presentation', was 'extremely articulate', and had presented a 'good analysis of the subject'. When asked to rate the talk, 85 per cent of the audience indicated that the incomprehensible Dr Fox had presented his material in a well-organized way, 70 per cent commended him on his use of examples, and almost 95 per cent found the talk stimulating.

Naftulin is not the only one to expose the human mind's wonderful ability to make something from nothing. In the mid-1960s MIT artificial intelligence expert Joseph Weizenbaum created a computer program that aimed to reproduce the experience of going to a psychotherapist. Named ELIZA (after Eliza Doolittle, the working-class flower seller who was taught how to speak proper in *Pygmalion*), the resulting program was very straightforward. At the start of each session, ELIZA asked people to state their problem, and then used a small number of simple rules to respond to different types of sentence. For example, if someone typed a comment with the structure 'I have just x', ELIZA would always reply 'You say you have just x?' Whenever the person said the words 'I think so', ELIZA responded with 'Do you say you think so for some special reason?' If someone's comment didn't match one of these standard forms, ELIZA was programmed to come out with a stock phrase, such as 'Tell me more.' So, imagine that you came to consult with ELIZA about a recent career change. The session might run something like this:

> ELIZA: Please state your problem.
> You: I have just changed jobs.
> ELIZA: You say you have just changed jobs?
> You: Yes, I am not sure I have made the right decision.
> ELIZA: You seem to be quite positive.
> You: I think so.
> ELIZA: Can you elaborate on that?
> You: Well, it feels like the right thing to do.
> ELIZA: I'm not sure I understand you fully.
> You: I am the sort of person that trusts my gut feelings.
> ELIZA: Your gut feelings?
> You: Yes, you know, when you have a hunch.
> ELIZA: You are sure.
> You: I think so.
> ELIZA: Do you say you think so for some special reason?
> You: Well, I do worry I was better off in my old job.

ELIZA: Were you really?
You: In some ways, yes.
ELIZA: I see.

Despite presenting people with a series of these completely meaningless statements, ELIZA proved extremely popular and convinced many people that they were indeed chatting to a genuine and experienced psychotherapist.

Similarly, politicians are also well aware of this effect, and often present the public with vague, ambiguous and even contradictory phrases, safe in the knowledge that voters will hear what they want to hear. ('We must be prepared to look back and have the courage to move forward, to acknowledge the rights of both workers and organizations alike, to support those in need without encouraging people to rely on the State'.) Even academics are not immune from the effect. In the mid-1990s physicist Alan Sokal from New York University thought that the same type of gobbledegook lay behind much postmodern cultural study, and decided to test his theory by submitting a completely meaningless article to an academic journal in the area.[11] The submission, entitled 'Transgressing the Boundaries: Towards a Transformative Hermeneutics of Quantum Gravity', consisted of irrelevant references, random quotations, and outright nonsense. For example, part of the article argued that quantum gravity had political implications and the piece concluded by noting: 'As yet no such emancipatory mathematics exists, and we can only speculate upon its eventual content. We can see hints of it in the multidimensional and non-linear logic of fuzzy systems theory; but this approach is still heavily marked by its origins in the crisis of late-capitalist production relations.'

The editors of the journal failed to identify the hoax and published the article.

This simple idea helps account for the success of psychic readings. Many of the comments made by psychics and mediums are ambiguous, and therefore open to several interpretations. When, for example, the psychic mentions picking up on 'a big change concerning property', they might be referring to moving house, helping someone else move house, inheriting a house, finding a new place to rent, or even buying an overseas holiday home. Because there is no timescale on the comment this move might have happened in the recent past, be happening right now or be going to happen in the near future. Clients work hard to make sense of such comments. They think back over their lives and try to find something that matches. In doing so, they can convince themselves that the psychic is very accurate. This process is often set in motion from the very start of the reading, with many psychics making it quite clear that they will not be able to deliver precise information. Instead, they claim that the process is like looking through smoked glass, or only just being able to hear voices in the darkness. It is up to the client to help out by filling in the gaps. Just like Dr Fox and ELIZA, the psychic then produces meaningless drivel that their clients transform into pearls of wisdom. Researcher Geoffrey Dean describes this phenomenon as 'The Procrustean Effect', after the mythical Greek figure who stretched or severed the limbs of his guests to ensure that they fitted into his bed.[12]

Mr D's readings were jammed full of such comments. Lisa was told that she was 'connected with something of a caring nature', that she was 'going through some sort of change in the workplace', that someone in her life was 'being especially difficult', and that she had recently received 'a gift from a young child'. One of the most dramatic moments in the reading came when Mr D told her that her brother had enjoyed a great deal of career success, and was considering joining an organization that would help him achieve even more. Mr D had no idea what he was talking about. His comment could, for example, have referred to Lisa's brother changing jobs, or becoming a member of a professional organization, gym, sports team, private club, or a trade union. However, Lisa's brother had recently been asked to join the Masons and so she interpreted Mr D's comments in that context. When we interviewed her afterwards, Lisa was especially impressed with this part of the reading, and misremembered Mr D's comments as explicitly referring to her brother and a Masonic Lodge.

So of the six psychological techniques that cold reading capitalizes on we have explored the 'Lake Wobegon' effect, the 'Dartmouth Indians versus the Princeton Tigers' effect, and the 'Doctor Fox' effect. Let's take a break before we look at the fourth key principle of cold reading . . .

HOW TO CONVINCE STRANGERS
THAT YOU KNOW ALL ABOUT THEM: PART ONE

Now it is time for you to master the psychological techniques used by professional psychics for your own wicked purposes. Before you start, decide what 'skill' you are going to appear to possess. It is best to choose something that appeals to the person that you are trying to impress. So if, for example, you think they are open to the idea of palmistry say that you can tell a great deal about them from the lines on their hand. If they are into astrology, explain that you can determine their past and future from their date of birth. Or, if they are sceptical about all matters paranormal, ask them to draw a picture of a house and use this as the basis for a 'psychological' reading.

Next, practise using the following three techniques:

1. Flattery

Start by telling them what they want to hear. Look at their palm, date of birth or drawing of a house, and explain that it reflects a very well-balanced personality. Try your best to keep a straight face as you appear to dig deeper, explaining that they appear to be terribly caring, responsible, friendly, creative, and polite. Also, don't forget to mention that they also seem to be highly intuitive, and so would be good at providing readings for others.

2. Double-headed statements

If you describe any trait and its exact opposite, people will only focus on the part of your description that makes sense to them. Work your way through the following five key personality dimensions using these double-headed sentences:

Openness: 'At times you can be imaginative and creative, but are more than capable of being practical and down-to-earth when necessary.'

Conscientiousness: 'You value a sense of routine in some aspects of your life, but at other times enjoy being spontaneous and unpredictable.'

Extroversion: 'You can be outgoing when you want to, but sometimes enjoy nothing better than a night in with a good book.'

Agreeableness: 'Your friends see you as trusting and friendly, but you do have a more competitive side that emerges from time to time.'

Neuroticism: 'Although you feel emotionally insecure and stressed, in general you are fairly relaxed and laid-back.'

3. Keep it vague

Although it is fine to drop in the odd specific statement ('Do you have a sister called Joanne, an irrational fear of porridge, and have you recently bought a yellow second-hand car . . . I thought not'), in general it is better to keep your comments vague. To help account for this vagueness, tell people that you sometimes struggle to understand the thoughts and images that cross your mind, and therefore they have to help you figure out what is going on.

In terms of actual statements, try, 'I am getting the impression of a significant change, perhaps a journey of some sort or an upheaval in the workplace', 'You have recently received a gift of some sort – perhaps money or something that has sentimental value?', 'I have a feeling that you are worried about a member of your family or a close friend?' Similarly, feel free to come out with abstract statements, such as 'I can see a circle closing – does that make sense to you?', 'I can see a door shutting – no matter how hard you pull it won't open' or 'I can see cleaning – are you trying to remove something or someone from your life?'

Now, let's resume our exploration of the principles of cold reading by going fishing.

4. Using the 'F Words'

During everyday conversations most people try their best to communicate their thoughts and opinions. However, even if just one person is speaking and another person is listening, information is not only flowing from the speaker to the listener. Instead, the conversation will always be a two-way process, with the listener constantly providing feedback to the speaker. Perhaps they will let the speaker know that they understand, and perhaps agree with, what is being said by nodding, smiling or saying 'yes'. Or maybe they will let the speaker know that they are confused, or don't agree with a comment, by looking confused, shaking their head, or saying 'you are a fool, please go away'. Either way, such feedback is vital to the success of our everyday conversations.

Psychics and mediums take this simple idea to the extreme. During a reading they will often make several comments, see which gets a reaction and elaborate on the selected option. Like a good politician or second-hand car salesman, they are not saying what is on their mind, but rather testing the water and then changing their message on the basis of the feedback they receive. This feedback can take many different forms.

They may look at whether their client nods, smiles, leans forward in their seat, or suddenly become tense, and alter their comments accordingly (this is one of the reasons that palmists are keen to hold your hand during a reading). The technique is referred to as 'fishing and forking', and Mr D was a master practitioner.

People tend to consult psychics about a relatively small number of potential problems, such as their health, relationships, travel plans, career, or finances. As Mr D worked his way through the Tarot cards he mentioned each of the topics and surreptitiously observed Lisa's reaction. She looked in good health and didn't really respond when he mentioned her having a few aches and pains. Questions about her career didn't produce much of a reaction. He then mentioned travelling but Lisa remained unmoved. Finally, he moved on to Lisa's emotional life. The moment he mentioned companionship Lisa's entire demeanour changed, and she suddenly looked very serious. Mr D immediately knew he was on to something and started to drill deeper. He looked at the lines on Lisa's palm, remarked on an imaginary blip in the heart line, and said that he wasn't certain whether it reflected a death in the family or a relationship that didn't work out. Lisa was completely unresponsive when he mentioned the death, but nodded as soon as she heard about the broken relationship. Mr D secretly noted the response and moved on. About ten minutes later he picked up another Tarot card and confidently announced that it showed that she had recently broken up with her partner. Lisa was stunned.

ARE YOU A GOOD JUDGE OF CHARACTER?

In addition to using the techniques described in this chapter, some cold readers say that they often get an intuitive feeling about a client, and these hunches have an uncanny knack of being accurate. What could account for these strange sensations, and are you a good judge of character?

A few years ago, psychologists Anthony Little from the University of Stirling and David Perrett from the University of St Andrews carried out a fascinating study into the relationship between people's faces and their personality.[13] The researchers had almost 200 people complete a personality questionnaire that measured each of the five dimensions described earlier in this chapter (openness, conscientiousness, extraversion, agreeableness and neuroticism). They then took photographs of those men and women who had the highest and lowest scores in each of the dimensions and used a computer program to blend each group of faces into a single 'composite' male and female image. They ended up with four separate composite images: one representing female low scorers, one representing female high scorers and vice versa for the male high and low scorers. The principle behind this technique is simple. Imagine having photographic portraits of two people. Both have bushy eyebrows and deep-set eyes, but one has a small nose while the other has a much larger nose. To create a composite of their two faces, researchers first scan both photographs into the computer, control for any differences in lighting, and then manipulate the images to ensure that key facial attributes – such as the corners of the mouth and eyes – are in roughly the same position. Next, one image is laid on top of the other, and an average of the two faces calculated. If both of the faces have bushy eyebrows and deep-set eyes, the resulting composite would also have these features. If one face has a small nose and the other has a large nose, the final image would have a medium sized nose.

The research team then presented these male and female composites to another group of 40 people and asked them to rate each of the faces on the different personality dimensions. Remarkably, their ratings were often highly accurate. For example, the composite image that had been made from the highly outgoing people was judged as especially extroverted, the composite created from the highly conscientious people was seen as particularly reliable, and so on. In short, your personality is, to some extent, written all over your face.

I have used the female composites created during this research to put together a quick and fun test to help discover whether you are a good judge of character.[14] To take part, simply answer the following five questions.

1. Which of these two composites looks the friendliest?

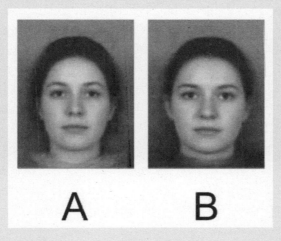

2. Which of these two composites looks the most reliable?

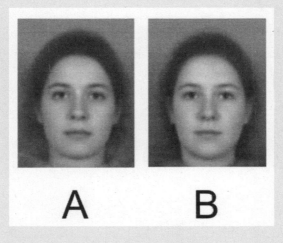

3. Which of these two composites looks the most outgoing?

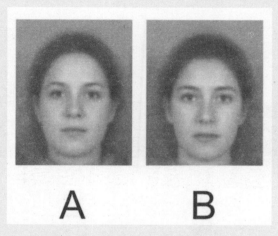

4. Which of these two composites looks the most anxious?

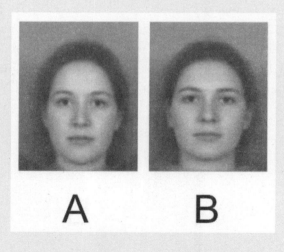

5. Which of these two composites looks the most imaginative?

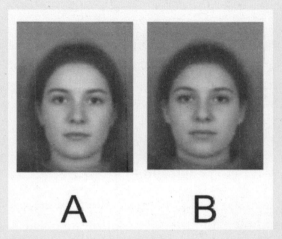

A   B

The correct answer to all of these questions is 'A'. How did you score? If you answered all of the questions correctly then it might be wise to trust your intuition about others. If you didn't, then you might be better off ignoring your hunches and finding out more about a person before making your mind up.

As we continue towards the fifth secret technique of psychics, I have a gut feeling that you are the sort of person that lets your heart rule your head, can sometimes be too impulsive for your own good, and have recently come into close contact with a goat. Rest assured you are not the only one.

5. The Illusion of Uniqueness

Towards the start of this chapter I asked you to carry out two simple psychological tests. One of them involved digging in a sandpit for buried treasure and the other involved thinking of one geometric shape inside another. Both of these give a vital insight into the fifth principle of cold reading.

I have asked many people to complete these two tasks. You might expect people to choose random locations in the sandpit. However, as shown in the plot opposite, the vast majority of them dig down in the same areas.

Similarly, when it comes to choosing two geometric shapes, most people tend to go with a circle inside a triangle, or vice versa.[15] However, the same sort of egocentric thinking that causes you to believe that you have an above average sense of humour and are more skilled than the average driver, also causes you to think that you are a unique and special individual. Although you might like to think that you are very different from other people, the truth of

the matter is that we are surprisingly similar and therefore remarkably predictable.

Psychics use this notion to give the impression that they have a paranormal insight into our personalities and past. Mr D explained that many psychics bolster their readings by using specific sounding statements that are likely to be true of many people. They might say that they have an impression of someone with a scar on their left knee (true for around a third of the population), own a copy of Handel's Water Music (again, about a third), have someone called 'Jack' in the family (true for one-fifth of people), have a key despite not knowing what it opens, or have a pair of shoes in the wardrobe that they know they will never wear again.[16] Mr D had developed several of his own over the years, including telling Lisa that he could see someone who was in need of medical care but were difficult to look after because they kept on throwing their pills down the sink, that someone in her family had once died without leaving a will, and that she had a stack of photographs in a drawer. Everyone assumes that they are unique, that these statements couldn't possibly be true of others, and so ends up being overly impressed.

It is now time to explore the sixth and final principle of cold reading. But before we do, let me make one final prediction. I have the impression that you arrange your books on the basis of the colour of their covers, and recently spent three days in Lisbon. No? That's not a problem.

6. Turning psychic lemons into lemonade

During his readings at Edinburgh University, none of Mr D's participants openly stated that any of his comments were untrue. However, it does happen. Under these circumstances psychics have various 'outs' to help them avoid outright failure. Perhaps the most common involves broadening a statement that has been rejected as incorrect. For example, 'I can sense someone called Jean' might be transformed into 'Well, if not Jean, perhaps Joan, or maybe even a Jack, but certainly a name starting with a J. Or something that sounds like a J. Like a K. Maybe Karen? Or Kate?'

There is also the strategy of giving someone else the problem by asking the person to think harder, or telling them that they may be able to work out the answer if they ask other members of their family after the reading. After that comes the old 'I was talking metaphorically' scam. Mr D told me that he had once been giving readings in a small seaside town. One of the readings was to a man named George. He looked at George's weather-worn face and guessed that he had spent much of his life outside, and had a hunch he might have worked on ships. Mr D looked at a Tarot card and said that he could see George standing by the port waiting for a ship to arrive. George looked disappointed and shook his head. He had spent his entire life working on a farm and didn't like the sea. It was a huge miss. In the blink of an eye Mr D explained that he was not talking literally but metaphorically. The ship was a new direction in George's life and he was nervous about the change. George's face lit up as he explained that yes, he had recently got married and was looking forward to sharing his life. Bingo. Psychic lemons transformed into lemonade in the blink of an eye.

HOW TO CONVINCE STRANGERS THAT YOU KNOW ALL ABOUT THEM: PART TWO

Earlier on we encountered three techniques essential to any psychic reading; flattery, double-headed comments and vague statements. Now it is time to learn how to use the three other techniques essential for a successful and convincing reading.

4. Fish and fork

It is important to touch upon a wide range of topics and then change your patter on the basis of the reaction you obtain. If your comment results in a blank look, play down the statement and move on to another topic. If you are greeted by nods and smiles, elaborate. Many manuals on how to give psychic readings recommend working through several key topics, with one writer recommending using the mnemonic 'THE SCAM' to remind yourself to cover Travel, Health, Expectations about the future, Sex, Career, Ambitions, Money.[17]

5. Predict the likely

Use statements that are true for many people, such as 'I can see you achieving something at school, perhaps getting an award – you can still remember feeling proud as the teacher called out your name', 'as a child, you had a particularly embarrassing experience that you still think about from time to time even today', 'why can I see the colour blue or purple? Are you thinking of buying something that colour, or have you just bought it?', 'who is the elderly woman I can see, someone in a black dress complaining about her legs?', and 'something happened about two years ago didn't it, a major change of some sort?'

Also, many psychic manuals advise readers to focus on the type of issues that concern people at different stages in their lives.[18] According to these writers, teenagers and those in their early twenties are often trying out different identities and exploring sexual relationships. Those in their mid-twenties to mid-thirties have usually developed a sense of stability and are more focused on their career, finances, and putting down roots. People in their mid-thirties to mid-forties are often worried about parental health and the stresses and strains of bringing up children. Those aged 45 onwards tend to be more concerned about their own health, whether their relationship is becoming stale, and the novelty of grandchildren.

6. Prepare your 'outs'

Remember that you cannot fail because if someone can't make sense of your statement you have two huge safety nets at your disposal. First of all, broaden your comment. Explain that perhaps the statement does not apply to the person directly, but rather someone in their family, a colleague at work or their friend. You can also broaden into the past and the future. Was this something that happened to them in childhood, or perhaps something that may take place in the near future? If that doesn't work, feel free to make the comment far more abstract explaining, for example, that when you see a 'holiday' you are actually referring to a major change of some sort, or when you speak about a 'hospital' you are really talking about someone coming into their life to care for them. If they are really struggling to make sense of a statement, make it sound as if they are the one that has failed by saying 'I'll leave that one with you'.

Finally, if all else fails, try using these giveaway signs to your advantage . . .

- Are the person's clothes slightly too small or big, or the surrounding buckle holes in their belt worn? If so, say that you have the impression that they have recently gained, or lost, some weight.

- Does their posture suggest time in the military, a background in dance, or perhaps someone who spends lots of time hunched up over their computer?

- Look at their skin and eyes. Dry-looking skin, and dull-looking eyes, are some of the most reliable signs of a long-term, or recent, health problem.

- Glance at their fingers and hands. Calloused hands suggest someone who is involved in manual work, while long nails on just one hand strongly suggest someone who plays guitar. Tar-stained fingers suggests a smoker, while a lighter coloured strip of skin on their ring finger will usually signal a recent break-up of a relationship.

- Shake hands with them. People with especially weak and limp handshakes tend to be more anxious than others.[19] Also, an unusually cold hand could be a sign that they suffer from poor blood circulation or are on some form of medication.

- Are their shoes practical or fashionable? Does this suggest that they are involved in sport, or vain? Also, especially large shoes might suggest that they work in a circus.

So there we have it. Mr D had lifted the lid off of the psychic industry. Learning how to give a psychic reading is not a question of attending psychic training courses or the school for gifted mediums. Instead, it is a case of flattery, double-headed statements, ambiguous comments, fishing and forking, predicting the likely, and transforming failure into success. It would be nice to think that Mr D was the only one engaged in fakery. Nice, but wrong. In fact there is an entire underground industry devoted to cold reading. Books with titles such as *Cashing in on the Psychic*, *Money-making Cold Reading*, and *Red Hot Cold Readings* are widely available; as are interactive DVDs, training courses and conventions all devoted to fooling all of the people all of the time.

Does this mean that all psychics and mediums are fakes? No. In fact, many more mediums and psychics are using the techniques described above without realizing it. Lamar Keene referred to them as 'shut-eyes' – people who don't have any paranormal ability but are, without being aware of it, fooling themselves and others.

Cold reading also explains why psychics have consistently failed scientific tests of their powers. By isolating them from their clients, psychics are unable to pick up information from the way those clients dress or behave. By presenting all of the volunteers involved in the test with all of the readings, they are prevented from attributing meaning to their own reading, and therefore can't identify it from readings made for others. As a result, the type of highly successful hit rate that psychics enjoy on a daily basis comes crashing down and the truth emerges – their success depends on a fascinating application of psychology and not the existence of paranormal abilities. Now that you know the techniques, going to a psychic or watching one on a television should be a very different experience. In the same way that a music lover appreciates the nuances of Mozart or Beethoven, so you will be listening out for the psychics fishing, broadening statements and forcing their clients to do their work for them.

Enjoy the concert.

2. OUT-OF-BODY EXPERIENCES

In which we hear about the scientists who attempted to photograph the soul, discover how a rubber hand reveals the truth about astral flying, learn how to leave our bodies and find out how our brains decide where we are right now.

I remember it as if it were yesterday. I had been admitted to hospital for some minor surgery and it was the night before the operation. As I drifted off to sleep something very strange happened. I felt myself slowly rise out of my bed, float up to the ceiling, and turned around to see my body sound asleep in the bed. A few seconds later I flew out of the door and whizzed full speed along the hospital corridors, eventually landing inside an operating theatre. The surgical staff were hard at work trying to remove a ketchup bottle from . . .

. . . actually, I can't carry on with the story. It's not that it is an especially painful memory, it's just that I feel bad for making the whole thing up. I have never had an out-of-body experience. Sorry for wasting your time – it's just that for years I have had to patiently listen to people as they describe their own flights of fancy and so it felt cathartic to produce one of my own.

Although fictitious, my experience does contain all of the elements associated with a 'genuine' out-of-body experience (or 'OBE'). During these episodes, people feel as if they have left their body and are able to fly around without it, with many being convinced that they have found out some information that they couldn't possibly have known otherwise. Many people report seeing their actual body during the experience, with some commenting on a strange kind of 'astral cord' linking their floaty self to their real self. Surveys suggest that between 10 and 20 per cent of the population has had an OBE, often when they are extremely relaxed, anaesthetized, undergoing some form of sensory deprivation such as being in a flotation tank, or on marijuana (bringing a new meaning to the term 'getting high').[1] If the experience occurs when a person is in a life-threatening situation, it can also involve the sensation of drifting down a tunnel, seeing a bright light, and a feeling of immense serenity (these tend to be referred to as near-death experiences, or 'NDE'). Experiences like these do seem to be surprisingly beneficial, with the vast majority of OBEers and NDEers reporting that the event has had a positive impact on their life.[2]

What is the explanation for these strange sensations? Are people's souls really drifting away from their earthbound bodies? Or are these light-headed moments the result of our brains playing tricks on us? And, if that is the case, what does that say about where we are the rest of the time?

Early attempts to answer these questions involved a small group of strange scientists joining forces with the grim reaper in a bizarre search for the soul.

Dead Weight

In 1861 Boston jeweller and keen amateur photographer William Mumler made a remarkable discovery.[3] As one of his self-portraits emerged from a developing tray he was astounded to see the ghostly figure of a young woman eerily floating by his side. Certain that the figure had not been present when he took the photograph, Mumler assumed that it was nothing more than a double exposure. However, when he showed the image to his friends they pointed out that the figure bore an uncanny resemblance to Mumler's dead cousin and became convinced that he had stumbled on a way of photographing the dead. Mumler's photograph quickly made front page news, with many journalists adopting a less than sceptical stance and promoting it as the first ever image of a spirit.

Sensing a business opportunity, Mumler promptly shut up his jewellery shop and started work as the world's first spirit photographer. In session after session he worked hard to ensure that the spirits appeared on cue, and soon the sound of his magnesium flash pot was matched only by the noise from his till. But after a few highly successful years, trouble started. Several eagle-eyed customers noticed that some of the alleged 'spirits' on their photographs looked remarkably like people who had attended Mumler's previous sittings. Other critics went further, accusing Mumler of breaking into houses, steal- ing photographs of the deceased and then using them to create his spirit images. The evidence stacked up and eventually Mumler was taken to court on charges of fraud. The trial proved a high-profile affair involving several well-known witnesses, including the famous showman Phineas Taylor 'there's a sucker born every minute' Barnum, who accused Mumler of taking advantage of the gullible (think 'kettle' and 'black'). Though acquitted of fraud, Mumler's reputation was ruined. Never recovering from the huge legal fees it had cost him to defend the case, he died in poverty in 1884.

Ironically, the notion of spirit photography survived Mumler's death. One eager proponent of the new fad was French researcher Dr Hyppolite Baraduc, who had a rather unusual take on the topic.[4] Well aware that many of the alleged spirits bore a remarkable resemblance to the living, and eager not to dismiss the entire enterprise as bunkum, Baraduc believed that the sitters were producing the images using their psychic powers. Excited by this thought, he conducted a series of studies in which he had people hold undeveloped photographic plates and concentrate on an image. When several of the plates revealed strange blobs and shapes, Baraduc rushed to the Paris Académie de Médecine and announced his findings.

Ignoring those who thought that his results were simply photographic artefacts, Baraduc forged ahead and started to experiment with other forms of supernatural photography. Although still sceptical of mainstream spirit photography, he wondered whether it might be possible to photograph the very recently deceased and capture the soul as it left the body. He was presented with his first opportunity to photograph the dead when his 19-year-old son Andre passed away from consumption in 1907. Just a few hours after Andre's death

Baraduc did what any loving father and dedicated scientist would have done – he snapped a picture of his son's lifeless body lying in its coffin and examined the resulting image for evidence of the soul. He was astounded to discover that the photograph showed a 'formless, misty, wave-like mass, radiating in all directions with considerable force'. Ignoring the possibility of this being some sort of photographic artefact, or indeed the result of him psychically projecting his own thoughts onto the image, Baraduc eagerly waited for another opportunity to test his hypothesis. He didn't have to wait long.

Just six months after the death of his son, Baraduc's wife became seriously ill and clearly did not have long to live. Eager to make the most of the opportunity, Baraduc set up his photographic equipment at his wife's bedside and patiently waited for her to shuffle off her mortal coil. His wife sighed three times as she passed away and Baraduc managed to take a photograph during one of her dying breaths. The image showed three luminous white 'globes' floating above Madame Baraduc. Elated, Baraduc took another photograph of his wife's corpse 15 minutes later and a third roughly an hour after that. The three mysterious globes made another appearance in the first of these images and congregated into a single large globe in the second.

Baraduc was certain that he had photographed the soul. Others were not so convinced. When assessing the images in his recent book *Ghosts Caught On Film*, Mel Willin notes that one professional photographer suggested that the effect could well have been caused by tiny pinholes in the bellows behind the lens of the camera.[5]

Baraduc was not the only soul-searching scientist to work

with the dying and dead. Just after the turn of the last century American physician Duncan MacDougall undertook a series of equally macabre, and now infamous studies in an attempt to discover the weight of the human soul.[6] He visited his local consumptives' home and identified six patients who were obviously very close to death (four from tuberculosis, one from diabetes, and one from unspecified causes). When each patient looked like they were just about to pop their clogs MacDougall quickly wheeled their beds onto an industrial-sized scale and waited for them to pass away. MacDougall's laboratory notes from one of the sessions provide a vivid description of the difficulties involved in the task:

> The patient . . . lost weight slowly at the rate of one ounce per hour due to evaporation of moisture in res-piration and evaporation of sweat. During all three hours and forty minutes I kept the beam end slightly above balance near the upper limiting bar in order to make the test more decisive if it should come. At the end of three hours and forty minutes he expired and suddenly, coincident with death, the beam end dropped with an audible stroke hitting against the lower limiting bar and remaining there with no re-bound. The loss was ascertained to be three-fourths of an ounce.

After another five patients had met their maker MacDougall calculated the average drop in weight at the moment of death, and proudly announced that the human soul weighed 21 grams. His findings guaranteed him a place in history and, perhaps more importantly, provided the title for a 2003 Hollywood blockbuster staring Sean Penn and Naomi Watts.

In a later study he dispatched 15 dogs on the scales and discovered no loss of weight, thus confirming his religious conviction that animals do not have souls.

When MacDougall's findings were published in the *New York Times* in 1907 fellow physician Augustus P. Clarke had a field day.[7] Clarke noted that at the time of death there is a sudden rise in body temperature due to the lungs no longer cooling the blood, and the subsequent rise in sweating could easily account for MacDougall's missing 21 grams. Clarke also pointed out that dogs do not have sweat glands (thus the endless panting) and so it is not surprising that their weight did not undergo a rapid change when they died. As a result, MacDougall's findings were confined to the large pile of scientific curiosities labelled 'almost certainly not true'.

A few years later American researcher Dr R. A. Watters conducted several remarkable experiments involving five grasshoppers, three frogs and two mice.[8] In 1894, Scottish physicist Charles Wilson was working on the summit of Ben Nevis when he experienced a 'Brocken spectre'. This striking optical effect occurs when the sun shines behind a climber and into a mist-filled ridge. In addition to creating a large shadow of the climber, the sunlight often diffracts through the water droplets in the mist, resulting in the giant figure being surrounded by coloured rings of light. The experience set off a chain of thought in Wilson that eventually resulted in him creating a device for detecting ionizing radiation known as a cloud chamber. Wilson's chamber consisted of a sealed glass container filled with water vapour. When an alpha or beta particle interacts with the vapour, they ionize it, resulting in visible trails that allow researchers to track the path of the particles.

It was the potential of the cloud chamber that enthralled Watters. In the early 1930s he speculated that the soul may have an 'intra-atomic quality' which might become visible if a living organism was exterminated inside Wilson's device. Watters didn't adopt Baraduc's 'keep it in the family' approach to research, or share MacDougall's scepticism about soulless animals, and so administered lethal doses of anaesthetic to various small creatures (including grasshoppers, frogs and mice), then quickly placed them into a modified cloud chamber. The resulting photographs of the dying animals did show cloud-like forms hovering above the victims' bodies. Even more impressive to Watters was the fact that the forms frequently seemed to resemble the animals themselves. Not only had he proved the existence of a spirit form, but he had also shown that frogs' souls, remarkably, are frog-shaped. His surviving photographs, now stored in the archives of the Society for Psychical Research in Cambridge, are less than convincing. Although the images do show large blobs of white mist, the shapes of the blobs would only resemble animals to those with the most vivid of imaginations. Once again, it is a case of the human mind seeing what it wants to see.

The ambiguous nature of the blobs proved the least of Watters' problems. Several critics complained that it was impossible to properly assess his spectacular claims because he had not described his apparatus in sufficient detail. Others argued that the images could have been due to him failing to remove dust particles from the chamber. The final nail in Watters' coffin came when a physics schoolteacher named Mr B. J. Hopper killed several animals in his own specially constructed cloud chamber and failed to observe any spiritual doubles.

The search for physical evidence of the soul proved less than impressive. Baraduc's mysterious white globes could well have been due to tiny holes in the bellows of his camera, MacDougall's loss of 21 grams at the moment of death was probably the result of idiosyncrasies in blood cooling, and Watters' photographs of animal spirits can be explained away as a combination of dust and wishful thinking. Given this spectacular series of failures it isn't surprising that scientists rapidly retreated from the photographing and weighing of dying humans and animals. However, reluctant to simply abandon the quest for the soul, they adopted an altogether different approach to the problem.

Anyone for tennis?

The Strange Case of the Spiritual Sneakers

Open nearly any New Age book about out-of-body and near-death experiences and you will soon read about Maria and the worn-out tennis shoe.

In April 1977 a migrant worker named Maria from Washington State suffered a severe heart attack and was rushed into Harborview Medical Centre. After three days in hospital Maria went into cardiac arrest, but was quickly resuscitated. Later that day she met with her social worker, Kimberly Clark, and explained that something deeply strange had happened during the second heart attack.[9]

Maria had undergone a classic out-of-body experience. As the medical staff worked to save her life, she found herself floating out of her body and looking down on the scene seeing a paper chart spewing out from a machine monitoring her vital signs. A few moments later she found herself outside the hospital looking at the surrounding roads, car parks and the outside of the building.

Maria told Clark that she had seen information that she could not have known from her bed, providing descriptions of the entrance to the emergency ward and the road around the hospital building. Although the information was correct, Clark was initially sceptical, assuming that Maria had unconsciously picked up the information when she had been

admitted to the hospital. However, it was Maria's next revelation that made Clark question her own scepticism.

Maria said that at one point on her ethereal journey she
had drifted over to the north side of the building, and that an
unusual object on the outside of a third floor window ledge
had caught her attention. Using her mind power to zoom in,
Maria saw that the object was actually a tennis shoe, and a
little more zooming revealed that the shoe was well worn and
the laces were tucked under the heel. Maria asked Clark if she
would mind seeing if the tennis shoe actually existed.

Clark walked outside the building and looked around,
but couldn't spot anything unusual. Then she went up to the
rooms in the north wing of the building and looked out of
the windows. Apparently this was easier said than done,
with the narrow windows meaning that she had to press her
face against the glass to see onto the ledges. After much face
pushing Clark was amazed to see that there was indeed an old
tennis shoe sitting on one of the ledges.

'Fifteen-love' to the believers.

As Clark reached out onto the ledge and retrieved the shoe
she noticed that it was indeed well worn and that the laces
were tucked under the heel.

'Thirty-love'.

Moreover, Clark noticed that the position of the laces
would only have been apparent to someone viewing the tennis
shoe from outside the building.

'Forty-love'.

Clark published Maria's remarkable story in 1985 and
since then the case has been cited in endless books, magazine
articles and websites as watertight evidence that the spirit can
leave the body.

In 1996 sceptic scientists Hayden Ebbern, Sean Mulligan and Barry Beyerstein from Simon Fraser University in Canada decided to investigate the story.[10] Two of the trio visited Harborview Medical Centre, interviewed Clark and located the window ledge that Maria had apparently seen all of those years before. They placed one of their own running shoes on the ledge, closed the window and stood back. Contrary to Clark's comments, they did not need to push their faces against the glass to see the shoe. In fact, the shoe was easily visible from within the room and could even have been spotted by a patient lying in a bed.

'Forty-fifteen'.

Next, the sceptics wandered outside the building and noticed that their experimental running shoe was surprisingly easy to spot from the hospital grounds. In fact, when they returned to the hospital one week later the shoe had been removed, further undermining the notion that it was difficult to spot.

'Forty-thirty'.

Ebbern, Mulligan and Beyerstein believe that Maria may have overheard a comment about the shoe while sedated or half-asleep during her three days in hospital, and then incorporated this information into her out-of-body experience. They also point out that Clark didn't publish her description of the incident until seven years after it happened, and thus there was plenty of time for it to have become exaggerated in the telling and retelling. Given that key aspects of the story were highly questionable, the trio thought that there was little reason to believe other aspects of the case, such as Maria saying that the shoe was well-worn prior to its discovery, and the lace being trapped under its heel.

'Deuce'.

Just a few hours at the hospital revealed that the report of Maria's infamous experience was not all that it was cracked up to be. Despite this, the story has been endlessly repeated by writers who either couldn't be bothered to check the facts, or were unwilling to present their readers with the more sceptical side of the story. Those who believed in the existence of the soul were going to have to come up with more compelling and water-tight evidence.

'New balls please.'

A QUICK VISUALIZATION EXERCISE

It is time for a simple two-part exercise. Both parts will require you to write in this book. You might be somewhat reluctant to do this, but it is important for three reasons. First, you will need to refer to the numbers later in this chapter and so it is helpful to have a permanent record of them. Second, if you are in a bookshop you will be morally obliged to buy the book. Third, if you have already bought the book, the chances of getting a decent resale price on eBay will be greatly diminished. OK, let's start.

Part One

Take a look at your surroundings. Perhaps you are in your home, lying in the park or sitting on the bus. Wherever, just have a look around. Now imagine how your surroundings would look if you were floating out of your body, about six feet above where you actually are, and looking down on yourself. Hold that image in your mind's eye. How clear is the image? If you had to assign it a number from one (where there is almost no image at all) to seven (a very clear and detailed image), what number would you give it? Now write down the number, in indelible blue or black ink, on the line below:

Your rating: _____

Now look around and see where you actually are, and then again imagine floating high above your body. Next, switch back to your actual location and then back to seeing the world from above your head. Now rate the ease with which you could switch between the two locations by coming up with a number between one ('Boy that was tricky') to seven ('Soooo easy'). Once again, write down the number below:

Your rating: _____

Part Two

Please rate the degree to which the following statements describe you by assigning each a number between one ('Absolutely not') and five ('Wow, it is like you have known me for years').[11]

RATING

1. While watching a film I feel as if I am taking part in it. ☐
2. I can remember past events in my life with such clarity that it is like living them again. ☐
3. I can get so absorbed in listening to music that I don't notice anything else. ☐
4. I believe that stoats work too hard. ☐
5. I like to look at the clouds and try to see shapes and faces in them. ☐
6. I often become absorbed in a good book and lose track of time. ☐

Many thanks for completing the exercises. More about them later.

How to Feel Like a Desk

The infamous case of the tennis shoe on the ledge provides less than compelling evidence for the notion that people are able to float away from their bodies. Worse still, several researchers invested a considerable amount of time and effort conducting more rigorous tests of the notion and also drew a blank. For example, parapsychologist Karlis Osis tested over a hundred people who claimed that they could induce an OBE at will, asking each to leave their body, travel to a distant room and identify the randomly selected picture that had been placed there.[12] The vast majority of his participants were confident that they had made the trip but as a group they scored no better than chance. Similarly, researcher John Palmer and his colleagues from the University of Virginia in Charlottesville used a variety of relaxation-based techniques to train people to have OBEs and then asked them to use their new-found ability to discover the identity of a distant target.[13] In a series of studies involving over 150 participants, the experimenters failed to detect any reliable evidence of extrasensory perception.

In short, over a hundred years of scientific soul searching has ended in failure. Despite Baraduc's attempts to photograph the spirits of his dead son and wife, MacDougall weighing the dying and Watters slaughtering several grasshoppers, the evidence didn't stack up. As a result, the researchers changed tack and focused their attention on the information provided by those who claimed to have left their bodies. The best anecdotal case studies turned out to be a tad unreliable, and experiments involving hundreds of OBEers attempting to identify thousands of hidden targets failed to yield convincing results.

After all of this, it might appear that out of the body experiences have nothing to offer the curious mind. However, subsequent work has adopted a very different approach to the problem and, in doing so, both solved the mystery and provided an important insight into the innermost workings of your brain.

There is an old joke about a man who is trying to track down a particular room in a University Philosophy Department. He becomes lost and eventually comes across a map of the building. On the map he sees a large red arrow pointing to a particular corridor, and on the arrow it says 'Are you here?' It's not a bad gag. But more importantly it raises an important issue – how do you know where you are? Or, to put it in slightly more philosophical language – why do you think that you are inside your own body?

It many ways, it seems like an odd question. After all, we seem to be inside our bodies and that is that. However, the question has hidden depths. Perhaps the greatest insights have come from a ground-breaking experiment that you can recreate in your own home using just a table, a large coffee-table book, a towel, a rubber hand and an open-minded friend.[14]

Start by sitting at the table and placing both of your arms on the tabletop. Next, move your right arm about six inches to the right and place the rubber hand where your right hand used to be (this is assuming that the dummy hand is a right hand – if not, use your left hand during the demonstration).

The set-up for the first part of the dummy hand experiment.

Now stand the book vertically on the tabletop between your right arm and the rubber hand, ensuring that it prevents you from seeing your right arm. Then use the towel to cover the space between your right hand and the rubber hand (see photograph below).

The dummy hand experiment in action. It is possible to appreciate the psychological impact of the experience by looking at the facial expression of the person in the photograph.

Finally, ask your friend to sit opposite you, extend their fingers and use them to stroke both your right hand and the rubber hand in the same place at the same time. After about a minute or so of stroking you will start to feel that the rubber hand is actually part of you. This feeling has interesting consequences for your real, but hidden, hand. Researchers have monitored the skin temperature of people's hands during the study and discovered that when they believe that the rubber hand is part of them, their hidden hand becomes about half a degree colder – it is as if the brain is cutting off the blood supply to the unseen hand once it believes that it is no longer part of the body.[15]

It is a powerful illusion. In a similar series of studies, conducted by Vilayanur Ramachandran and described in his book *Phantoms in the Brain*, people were asked to place their left hand below a table, and an experimenter then stroked the hidden hand and the tabletop simultaneously.[16] Once again, their sense of self shifted, with about 50 per cent of people feeling as if the wooden tabletop had become part of them.

To explain what's going on here, let's use a simple analogy. Imagine walking around in a new city and suddenly realizing that you are lost. The only way forward is to go hunting for a signpost. Similarly, when your brain is trying to decide where 'you' are it has to rely on the equivalent of signposts, namely, information from your senses.

Most of the time this works really well. Your brain might, for example, see your hand and feel pressure from your fingertip, and so correctly assume that 'you' are in your arm. However, in the same way that people sometimes mess around with signposts and point them in the wrong direction, so once in a while your brain will mess up. The rubber hand experiment is one of those situations. During the study, your brain 'feels' your left hand being stroked, 'sees' a dummy hand or wooden table being subjected to simultaneous stroking, concludes that 'you' must therefore be located in the dummy hand or table, and constructs a sense of self that is consistent with this idea. In short, the sense of where you are is not hard-wired into your brain. Instead, it is the result of your brain constantly using information from your senses to come up with a sensible guess. Because of this, the sense of 'you' being inside your body is subject to change at a moment's notice.

Ramachandran's work has important practical, as well as theoretical, implications. The majority of people who have had an arm or leg amputated often continue to feel excruciating levels of pain from their phantom limb. Ramachandran wondered whether this pain was due, in part, to their brains becoming disoriented because they were continuing to send signals to move the missing limb but then not seeing the expected movement. To test his theory, Ramachandran and his colleagues ran an unusual experiment with a group of amputees who had lost an arm.[17] The research team built a two-foot-square cardboard box that was open on the top and front. They then placed a vertical mirror along the middle of the box, thus separating it into two compartments. Each participant was asked to place their arm into one of the compartments, and then orient themselves so that they could see a reflection of their arm in the mirror. From the amputee's perspective it appeared as if they were seeing both their actual, and missing, arm. The amputee was then asked to carry out a simple movement with both of their hands at the same time, such as clenching their fists or wriggling their fingers. In short, Ramachandran's box created the illusion of movement in their missing limb. Amazingly, the majority of the participants reported a reduction in the pain associated with their phantom limb, with some of them even asking if they could take the box home with them.

It is one thing to convince people that part of them inhabits a dummy hand or tabletop, but is it possible to use the same idea to move a person out of their entire body? Neuroscientist Bigna Lenggenhager, from the École Polytechnique Fédérale de Lausanne in Switzerland, decided to investigate.[18]

If you were to take part in one of Lenggenhager's studies you would be taken into her laboratory, asked to stand in the centre of the room and fitted with a pair of virtual reality goggles. A researcher would then place a camera a few feet behind you and feed the output into your goggles, causing you to see an image of your own back standing a few feet in front of you. Next an animated stick would appear on the image in front of you and slowly stroke your virtual back. At the same time the researchers would sneak up behind the real you and slowly stroke your back with a highlighter pen, being careful to ensure that the actual stroking matched the virtual stroking. The experimental set-up is identical to the dummy hand study, but with the 'virtual you' taking the place of the dummy hand and a highlighter pen replacing your friend's hand. In the same way that stroking the dummy hand produced the strange sensation that part of you inhabited the hand, so Lenggenhager's set-up resulted in people feeling as if their entire body was actually standing a few feet in front of themselves.

The dummy hand and virtual reality experiments demonstrate that the everyday feeling of being inside your body is constructed by the brain from sensory information. Alter that information and it is relatively easy to get people to feel as if they are outside of their bodies. Of course, people don't have access to rubber hands and aren't wired into virtual reality systems when they have out-of-body experiences. However, many researchers now think that this strangely counter-intuitive idea is essential to understanding the nature of these episodes.

MIRROR, MIRROR ON THE WALL

Neuroscientist Vilayanur Ramachandran and his colleagues have created a simple way of replicating Lenggenhager's experiment without the need for a complicated and expensive virtual reality system.[19] In fact, you just need two large mirrors and your finger. Arrange the two mirrors so that they are facing one another and a few feet apart. Next, angle one of the mirrors so that when you look into one mirror you see the reflection of the back of your head (see photograph). Finally, gently stroke your cheek with your finger and look at the image in the mirror.

Set-up for the mirror experiment.

This rather unusual set-up replicates the illusion created by Lenggenhager's virtual reality system. Your brain 'feels' your cheek being stroked, 'sees' a person standing in front of you being subjected to simultaneous stroking, concludes that 'you' must therefore be standing there, and constructs a sense of self that is consistent with this idea.

When he took part in the demonstration, Ramachandran felt as if he was touching an alien or android body that was outside his own body. Many of his colleagues felt similar sensations, with some of them reporting that they wanted to say 'hello' to the person in the mirror.

At the start of this book I described how seeing psychologist Sue Blackmore on television made me realize how studying the supernatural could reveal important insights into our brains, behaviour and beliefs. Blackmore has investigated many aspects of the paranormal over the years, but much of her work has focused on the secret science behind out-of-body experiences.

Witchcraft, LSD and Tarot Cards

Sue Blackmore's interest in the paranormal dates back to 1970 when she was a student at Oxford University and had a dramatic out-of-body experience. After several hours experimenting with a Ouija board and then relaxing with some marijuana, Sue felt herself rise out of her body, float up to the ceiling, fly across England, travel over the Atlantic, and hover around New York. Eventually she travelled back to Oxford, entered her body through her neck and finally expanded to fill the entire universe. Other than that it was a quiet night.

Upon her return to reality, Sue became fascinated with weird experiences, trained as a white witch, and eventually decided to devote herself to parapsychology. She was awarded a doctorate for work examining whether children have telepathic powers (they didn't), went on several LSD trips to see if they would improve her psychic ability (they didn't), and learned to read Tarot to discover if the cards could predict the future (they didn't). After 25 years of such disappointing results Sue finally gave up the ghost and became a sceptic. For many years she examined the psychology of paranormal experiences and beliefs, trying to figure out why people experienced seemingly supernatural sensations and bought into such strange stuff. Most recently she has turned her attention to the mystery of consciousness, focusing on the ways in which the brain creates a sense of self (although, rather disappointingly, the 'Who Am I' tab on her website delivers a straight biography).

One of Blackmore's early investigations tackled a question that comes up frequently when I speak about the paranormal – why do identical twins often appear to have a strange psychic bond with one another? Many proponents of psychic ability believe that this odd bond is due to telepathy. In contrast, sceptics argue that twins will often think in very similar ways because they have been raised in the same environment and have the same genetic makeup, and that such similarity will cause to them to make the same decisions and thus appear to read each other's minds.

To help settle the issue, Blackmore brought together six sets of twins and six pairs of siblings, and conducted a two-part experiment.[20] The first part was a straightforward test of telepathy. One member of each pair played the role of the 'sender' while the other was the 'receiver'. The sender was presented with various randomly selected stimuli (such as a number between one and ten, an object, or a photograph), and was asked to psychically transmit the information to the receiver. No evidence of telepathy emerged from either the twins or the siblings.

In the second part of the experiment, Blackmore asked the senders to transmit the first number that came into their mind, make any drawing that appealed to them, and choose which of four photographs to send. The results suddenly changed. As predicted by the 'twin telepathy is due to similarity' hypothesis, there was a sudden surge in the twins' performance. For example, when asked to think of a number between one and ten, 20 per cent of the trials involving twins produced the same number compared to just 5 per cent of those with the siblings. For the drawings, the twins again scored well, exhibiting a 21 per cent success rate compared to the siblings' 8 per cent.

In short, the evidence indicates that twin telepathy is due to the highly similar ways in which they think and behave, and not extra-sensory perception.

Interview with Sue Blackmore
www.richardwiseman.com/paranormality/SueBlackmore.html

However, Blackmore is perhaps best known in sceptical circles for her work explaining out-of-body experiences. She took as her starting point the notion that the feeling of being located inside your body is an illusion created by your brain on the basis of incoming sensory information. Then, in the same way that a rather weird set of circumstances involving a dummy hand or a virtual reality system can cause people to believe that they are elsewhere, Blackmore wondered whether an equally strange set of circumstances might cause people to think that they had floated away from their bodies. Sue focused her attention on two elements that were central to most OBEs.

The first principle can be illustrated with the help of the image over the page.

Fix your eyes on the black dot in the centre of the image and stare at it. Providing that you are able to keep your eyes and head relatively still you will find that after about 30 seconds or so the grey area around the dot will slowly fade away. Move your head or eyes and it will jump right back again. What is going on here? It is all about a phenomenon referred to as 'sensory habituation'. Present someone with a constant sound, image, or smell and something very peculiar happens. They slowly get more and more used to it, until eventually it vanishes from their awareness. For example, if you walk into a room that smells of freshly ground coffee, you quickly detect the rather pleasant aroma. However, stay in the room for a few minutes, and the smell will seem to disappear. In fact, the only way to re-awaken it is to walk out of the room and back in again. In the case of the illustration above your eyes slowly became blind to the grey area because it was unchanging. This exact same concept can result in the so-called 'hedonistic treadmill', with people quickly getting used to their new house or car, and feeling the need to buy an even bigger house or better car.

Blackmore speculated that this process was also central to OBEs. People tend to experience OBEs when they are in situations in which their brains are receiving a small amount of unchanging information from the senses. They are often robbed of any visual information because they have their eyes shut or are in the dark. In addition, they usually don't have any tactile information because they are lying in bed, relaxing in the bath, or are on certain drugs. Under these circumstances the brain quickly becomes 'blind' to the small amount of information that is coming in, and so struggles to produce a coherent image of where 'you' are.

Like nature, brains abhor a vacuum, and so start to generate imagery about where they are and what they are doing. That is part of the reason why people are more likely to have images flowing through their mind when they shut their eyes, are in the dark or take drugs. Blackmore hypothesized that certain types of people would naturally find it easy to imagine what the world looks like when you float out of your body, and also become so absorbed in their imagery that they confuse imagination for reality, and that these individuals that would be especially likely to experience OBEs.

To test her theory Blackmore carried out several experiments.[21] In fact, you have already taken part in a version of them. A few pages ago I asked you to imagine yourself being about six feet above where you actually are, and rate the clarity of your imagery and the ease with which you switched from one perspective to another. Sue presented this task to two groups of people – those that had experienced an out-of-body experience and those that had not – and obtained very different results. Those that had previously experienced floating away from themselves tended to report much more vivid images and found it much easier to switch between the two perspectives.

Blackmore also speculated that people who reported OBEs would tend to become absorbed in their experiences, so that they found it difficult to separate fact from fantasy. I also asked you to rate the degree to which six statements described you. Five of them are the types of items that you find on standard questionnaires designed to measure the degree to which you become absorbed in your experiences (I added the item about the stoats working too hard for fun). People who obtain high scores on absorption questionnaires tend to lose track of time when they watch films and television programmes, become confused about whether they have actually carried out an action or simply imagined it, and are more easily hypnotized (in the case of the five questions presented at the start of this chapter a total of 20 or more would constitute a high score). In contrast, lower scorers are more down-to-earth, practical and rarely confuse their imagination with reality (a low score would be ten or less). Blackmore's studies involved asking OBEers and non-OBEers to complete absorption questionnaires: the OBEers consistently obtained much higher scores.

In short, Blackmore's data suggests that people who experience OBEs are much better than others at naturally generating the type of imagery associated with the experience, and struggle to tell the difference between reality and imagination. Put these people in a situation where their bodies receive only a small amount of unchanging information about where they actually are and, just like the people taking part in the dummy hand and virtual reality experiments, they can end up believing that they are no longer located inside their bodies.

HOW TO LEAVE YOUR BODY

Understanding the real causes of out-of-body experiences can help you become a frequent flyer. The first part of the process involves developing three key psychological skills: relaxation, visualization and concentration. Let's examine each in turn.

Relaxation

'Progressive Muscle Relaxation' involves deliberately tensing various muscle groups and then releasing the tension. To try the technique, remove your shoes, loosen any tight clothing and sit in a comfortable chair in a quiet room. Focus your attention on your right foot. Gently inhale and clench the muscles in your foot as hard as possible for about five seconds. Next, exhale and release all of the tension, allowing the muscles to become loose and limp. Work your way around your body performing the procedure in the following order:

1. Right foot	7. Right hand	13. Abdomen
2. Right lower leg	8. Right forearm	14. Chest
3. Entire right leg	9. Entire right arm	15. Neck and
4. Left foot	10. Left hand	shoulders
5. Left lower leg	11. Left forearm	16. Face
6. Entire left leg	12. Entire left arm	

Each time, tense the appropriate body part for about five seconds and then release the tension.

Visualization

Inducing an out-of-body experience requires good visualization skills. If you are naturally good at imagining scenes and pictures in your head then that is great. If not, try the following exercise.

Imagine walking into your kitchen, taking an orange out of the cupboard and placing it on a green plate. Next, think about digging your nails into the smooth skin of the orange and starting to peel it. Think about how the orange would feel and smell. Visualize the juice coming out of the orange and onto your fingers. Imagine pulling all of the peel away and placing it on the plate. In your mind's eye, separate each of the segments and place them on the plate as well. Now look at the juicy segments. Are you salivating? Are the colours bright and sharp? Was each stage of the process vivid and did it involve all of your senses?

Repeat the exercise once every few days, trying to make it seem more realistic each time.

Concentration

The ability to focus your thoughts is also key to creating an out-of-body experience. This simple exercise will help assess and, if necessary, improve your concentration skills.

Try to count from 1 to 20 in your mind, moving onto each new number after a few seconds. However, the moment that any other thought or image enters your mind, start the count again. Initially, you will probably find this simple task surprisingly difficult, but over time you will learn how to focus your thoughts and will soon find yourself counting to 20 with no distractions.

Putting it all together

OK, it's time to try and induce an out-of-body experience. Sit in the most comfortable chair in your house. Next, stand up and take a look around. How does the room look from this perspective? Remember as many details as possible including, for example, the position of any furniture, the scene outside the window, and any pictures on the walls. Next, slowly make your way to another room. Once again, notice as much as possible on the way, including the colour of the walls, the furniture and objects that you encounter, and the type of flooring that you are walking on. To help with the process, choose four key points along the route and remember them in as much detail as possible.

Now return to the original room and sit down in the chair. Carry out the 'Progressive Relaxation Exercise'. When you feel completely relaxed, imagine a duplicate of yourself standing in front of you. To avoid the difficult (and for many, unpleasant) task of visualizing your face, imagine that your doppelganger is standing with their back to you. Try to form an image of their clothes and the way they are standing. Now, think back to what you saw when you were actually standing in that position, and imagine moving from your body into theirs. Don't worry if you don't succeed at first. This is tricky stuff and usually requires some practice.

Once you manage to feel as though you have left your own body and entered the mind of your imaginary doppelganger, try to take a few steps along the route that you mapped out, stopping at each of the four points to admire the view. If you are struggling with the movement, some researchers recommend increasing your motivation by not drinking any liquid for a few hours before the experience, and placing a glass of water in the room that you intend to visit. Also, don't be afraid of the experience – remember, you can snap back into your body at any point. After you have got the hang of inducing an out-of-body experience, you should be able to fly around the world at will, limited only by your imagination and without feeling guilty about your carbon footprint.

For decades a small number of devoted scientists attempted to prove that the soul is able to leave the body. They took photographs of recently deceased family members, weighed the dying, and asked those having out-of-body experiences to try to see pictures hidden in distant locations. The enterprise failed because you are a product of your brain and so cannot exist outside of your skull. Subsequent research into out-of-body experiences focused on finding a psychological explanation for these strange sensations. This work revealed that your brain constantly relies on information from your senses to construct the feeling that you are inside your body. Fool your senses with the help of rubber hands and virtual reality systems, and suddenly you can feel as if you are part of a table or standing a few feet in front of your body. Rob your brain of these signals and it has no idea where you are. Couple this sense of being lost with vivid imagery of flying around, and your brain convinces itself that you are floating away from your body.

Your brain automatically and unconsciously carries out the vitally important 'where am I?' task every moment of your waking life. Without it, you would feel that you are part of the chair you are sitting on one moment and in the floor the next. With it, you have the stable sense of constantly being inside your body. Out-of-body experiences are not paranormal and do not provide evidence for the soul. Instead, they reveal something far more remarkable about the everyday workings of your brain and body.

3. MIND OVER MATTER

In which we discover how one man fooled the world,
learn how to bend metal with the power of our minds,
investigate gurus in India and find out why we
sometimes cannot see what is happening
right in front of our eyes.

Born in New Jersey in 1959, James Alan Hydrick had a tough childhood.[1] When he was three years old his alcoholic mother ran away from the family, leaving her equally alcoholic husband to bring up Hydrick on his own. When Hydrick was six years old a bad situation became even worse when his father was convicted of armed robbery and sentenced to two years in prison. This, combined with rumours that Hydrick was a victim of physical abuse, caused social services to move him into foster care. Unfortunately, Hydrick's behaviour proved problematic, and he was moved from one foster family to another.

When he was eighteen, he was convicted of kidnapping and robbery, and spent time in the Los Angeles County Jail. While behind bars, he developed an avid interest in martial arts and worked hard to master various fighting techniques. Around the same time he also appeared to manifest psychokinetic powers. In what was to become his best-known demonstration, Hydrick would balance a pencil lengthways over the edge of a table and 'will' it to move. With his head turned in the opposite direction and hands away from the table, the pencil would slowly rotate, then stop and reverse direction. On other occasions he would open the prison Bible and ask Jesus to make his presence known. The pages of the good book would turn over one after another as if being turned by a ghostly hand.

When he was released from prison, Hydrick travelled to Salt Lake City, set up the 'Institute of Shaolin Gung Fu', and offered to help others learn martial arts and develop their psychokinetic abilities. In addition to moving pencils and fluttering the pages of Bibles, Hydrick added other stunts to his psychic repertoire, including making heavy punch bags in his Institute's gym swing without touching them.

In December 1980 he was invited to demonstrate his powers on ABC's TV programme *That's Incredible!*. Each week the show featured a bizarre mix of stunts and performers, including a record-breaking sword-swallower, a group of rats that played basketball in a specially constructed mini-court, and a man who was prepared to be dragged along the ground on a metal tray at over a hundred miles per hour. The programme attracted a huge audience and represented a golden opportunity for Hydrick to hit the big time.

Hydrick (who by this time had adopted the mysterious-sounding stage name 'Song Chai') opened the segment by performing his psychokinetic page-turning stunt. All went well, with the studio audience shouting 'That's Incredible!' on cue, and the phrase appearing in large block capitals across the screen for the hard of thinking. He then chatted about his abilities with the hosts and performed the pencil stunt. The audience were impressed.

Then it happened. Host John Davidson, who was sitting closest to Hydrick during the pencil demonstration, said that he thought he had heard Hydrick blowing on the pencil. Hydrick looked hurt and denied the accusation. A dramatic hush spread over the audience, presumably as they readied themselves to shout 'Actually, If That Is The Case, That's Not Quite So Incredible!' With his back against the wall, Hydrick turned to Davidson and asked, 'Do you want to put your hand over my mouth?' Davidson agreed, and the studio audience held their breath as Hydrick focused on moving the pencil. A few seconds later the pencil slowly rotated around. Davidson looked stunned and the audience went wild.

Word of Hydrick's remarkable abilities quickly spread, with one national tabloid going so far as to label him 'The World's Top Psychic'. He seemed destined for a place in the psychic hall of fame. And he might well have achieved it if it hadn't been for James 'The Amazing' Randi.

That's My Line

In chapter one we learned how magician and arch-sceptic James Randi has devoted his life to paranormal myth-busting, offering a million dollars to anyone who can demonstrate the existence of paranormal abilities under scientifically controlled conditions (his money remains unclaimed).

Hydrick's show-stopping demonstrations on *That's Incredible!* caught Randi's eye and he challenged the young psychic to perform his feats under more controlled conditions. In February 1981, the two of them crossed swords on another light entertainment television programme called *That's My Line*. At the start of the segment host Bob Barker introduced Hydrick and asked him how he had developed his psychic powers. Hydrick seemed to forget about his time behind bars, explaining that a wise old Chinese man called Master Wu had taught him how to reach the fourth level of consciousness (which, it seems, also involved the ability to be highly economical with the truth about his alleged psychic powers). Hydrick then demonstrated his amazing pencil-moving abilities and the audience applauded. Next, Barker placed an open telephone directory on the table and Hydrick called upon the great operator in the sky to help turn the pages. After several aborted attempts, and 25 minutes of less than riveting television, he caused a page in the book to flip over.

In the second part of the segment, Barker introduced Randi, who unlocked a large trunk at the back of the stage and removed his secret weapon – a tube of Styrofoam chips. Randi scattered the chips all around the open telephone book and challenged Hydrick to again turn over one of the pages using the power of his mind. Randi explained that he suspected Hydrick had been turning the pages by secretly blowing on them and that if he tried this again the Styrofoam chips would go flying.

Under the watchful eyes of three independent scientific experts Hydrick tried to move a page. After 40 minutes of hand-waving and brow-furrowing, and with the audience getting increasingly hungry and restless, he admitted defeat. According to Hydrick the Styrofoam chips and studio lights were forming static electricity which was pulling the page down and disrupting his psychic performance. Both Randi and the panel of experts agreed that this sounded like total tosh. Hydrick was adamant that his feats were not due to trickery and again tried to move the page using his psycho-kinetic powers. Once again, his abilities deserted him. Barker, Randi and the independent panel gave Hydrick the thumbs down, and the audience finally got to eat.

Hydrick's appearance on *That's My Line* was not a great career move. Although his most devout supporters might have been able to convince themselves that their hero was simply unnerved by the sudden introduction of sceptical observers and Styrofoam chips, most viewers came away with the distinct impression that Hydrick's line was one of trickery. He knew that he needed a saviour. A man who could both promote his abilities and cleanse his public soul of alleged deception. Enter the third and final character in the story – Danny Korem: one-time magician, psychic investigator and self-professed Messianic Jew.

Psychic Puffery

Nowadays, Danny Korem is president of Korem & Associates, a company specializing in 'rapid-fire on-the-spot behavioural profiling'. According to the company's website, their unique training programme can help people accurately judge the motivation, personality and communication style of others in seconds. However, in the 1980s Korem was leading a somewhat different life.

Korem had gained a considerable reputation as a skilled magician and had, according to his current online résumé, 'read or reviewed over 10,000 books, manuscripts, and periodicals on deception'. He had also developed a keen interest in the paranormal and, like Randi, had written extensively about the tricks of the psychic trade. However, unlike Randi, Korem was a strong believer in God and had co-authored a book, entitled *Fakers*, to help people separate fake supernatural phenomena from the real stuff. (In one section, Korem writes: 'As stated in Chapter 10, spirits of the dead cannot come back, because of the spiritual laws the Lord has set up'[2].)

In the first part of this genuinely informative but deeply odd book, Korem explains the psychological basis for lots of seemingly paranormal phenomena, including table-tipping, the Ouija board and fire-walking. In the second part he discusses 'genuine' supernatural phenomena, explaining, for example, that demons are scattered over the face of the planet and so may appear to be able to predict the future by drawing information from a great many sources ('Angels were never given this power'). On a more down-to-earth level, Korem also offers practical advice to those trying to separate cases of genuine possession from instances in which a person requires psychiatric care (as Korem notes, 'The key word is *balance*').

Korem became fascinated with Hydrick and arranged to meet him. He decided not to tell Hydrick about his background in magic ('Thou shalt not bear false witness'), and instead posed as a documentary maker keen to shoot a film about Hydrick's life and powers. No doubt eager to recover from the damage inflicted by his appearance on *That's My Line*, Hydrick agreed to be involved. After carefully observing Hydrick performing his page-turning and pencil-moving demonstrations, Korem became convinced that Randi was right: Hydrick was not using any form of psychokinetic ability but rather blowing on the objects in a highly deceptive and skilful way. Instead of confronting him straight away, Korem returned home and worked hard to duplicate all of Hydrick's methods ('Thou shalt not steal'). After much huffing and puffing, Korem felt ready to move onto the next stage of his cunning plan.

Korem asked whether he could film Hydrick exhibiting his powers. Hydrick agreed, and happily came along to a taping session and demonstrated his pencil-moving and page-turning abilities. Hydrick was then asked whether he would mind trying to transfer his remarkable powers to Korem. This was not a novel request to Hydrick. In fact, he frequently told people that he could bring out their latent psychic powers, and would then blow as the person moved their hands around the object, thus giving the impression that they did have abilities. Hydrick placed his hands above Korem's hands and concentrated for a few moments. Korem then leant forward and used the power of breath to move a pencil. Hydrick looked confused and stunned.

Korem then arranged to interview Hydrick. Taking his life in his hands, Korem told the martial arts expert that he had figured out his methods and that the game was up. Hydrick calmly confessed all. He explained that as a nine-year-old he had seen an American magician named Harry Blackstone Junior, and became fascinated by the psychology of deception. Around the same time his father repeatedly locked him in a cupboard as a punishment for bad behaviour, and so he created the imaginary Master Wu to keep him company. Hydrick went on to admit that Korem and Randi were right – all of his supposedly psychokinetic demonstrations were achieved by air currents. (The only exception was the movement of the punchbags – that was due to them hanging from a metal roof that expanded under the heat of the sun.) Towards the end of the interview Korem asked Hydrick why he had felt the need to fake psychic powers. Hydrick explained that he longed for the attention that he didn't receive as a child and, after a lifetime of being told that he was stupid, wanted to show that he was capable of fooling the world.

Soon after taping his confession, Hydrick was arrested for breaking and entering. He then escaped from confinement, was re-arrested, escaped again, and was re-re-arrested. Upon his release from jail in late 1988 he moved to California and soon attracted the attention of the police when he started to use his psychic stunts to befriend a group of young boys. When evidence of molestation emerged the police took action and issued a warrant for his arrest.[3] Hydrick fled, but then accepted an invitation to appear on a national television show and was subsequently recognized by an off-duty Californian police officer. Hydrick was re-arrested. Still unable to shake off his reputation as a psychic, the security guards driving Hydrick back to California became anxious when they thought he was using his supernatural powers to rock the van, and later warned prison staff not to look him straight in the eye because he might cast a spell on them. A few months after his arrest Hydrick was convicted on several counts of child molestation and sentenced to 17 years in prison.

In 2002 a British television programme listed the 50 greatest magic tricks in the world. Hydrick's pencil and page-moving demonstrations came in 34th, beating Uri Geller's alleged metal-bending by five places.

THE PINOCCHIO TEST

Fake psychics possess an innate ability to deceive others. Take this simple test to discover whether you too are a natural-born liar.[4]

Imagine sitting across a table from a friend. The following four cards are lying face up on the table in front of the two of you, but there is a barrier in front of one of the cards (in this case the one with the triangle on it) so that your friend can't see it but you can.

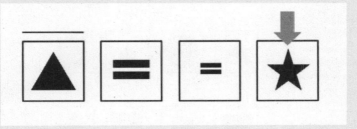

The aim is to speak to your friend and tell them to pick up the card with the star on it (shown by the arrow), but without giving away any information about the hidden symbol. You are not allowed to mention the position of the card so you might say something like 'Please pick up the card with the star on it', and your friend would lean forward and pick up the correct card. Got the idea? OK, turn over the page and try the following five sets of cards.

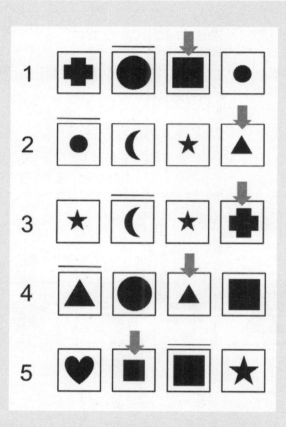

Finished? The test revolves around your behaviour on the fourth and fifth sets of cards. Good deceivers naturally think about how any situation looks from another person's point of view. On the fourth trial you were shown a small triangle and were asked to conceal the large triangle. However, from your friend's perspective there was only one triangle – the small one. Therefore, if you said

'Please pick up the card with the small triangle on it', this will give your friend a clue that the hidden card has a large triangle on it. How did you perform? The same applies to the final set of cards. Did you ask your friend to pick up the card with the 'square' or the 'small square'?

Try the test on your friends, colleagues and family to identify those with naturally deceptive tendencies!

Fooling all of the People all of the Time

Magicians and fake psychics consistently deceive one of the most sophisticated, complex, and impressive evolutionary triumphs in the world – the human brain. They face a formidable foe. Brains have put mankind on the moon, helped rid the world of major diseases, and unravelled the origins of the universe. How then do people like Hydrick deceive these finely-honed thinking machines?

Most magicians believe that the answer lies in their secret knowledge about how to fake the impossible, and so are fiercely secretive about their methods. However, as illusionist Jim Steinmeyer so eloquently put it in his book *Art & Artifice and Other Essays on Illusion*, they are guarding an empty safe.[5] In the same way that Hydrick blew on the objects in front of him, so the methods employed by magicians often amount to little more than sleight of hand, a rubber band or a concealed trapdoor. The real secrets of magic are psychological, not physical. Like most fake psychics, Hydrick employed five different psychological principles to transform the act of blowing into an alleged miracle. Each principle is designed to act like a wall that prevents people from entering the performer's inner sanctum and finding out what is really going on. Understand the principles and you will understand how Hydrick and others have fooled the world.

The first all-important issue is selling the duck.

Selling the Duck

Imagine that you really like ducks. In fact, you don't just like them, you adore them. You love the shape of their beaks, the silly 'quack' noise they make, you'd love a pet duck and you think it's cute the way your friends quickly lower their heads whenever you mention them. Now imagine that someone shows you the picture below.

It would not be at all surprising if you see a duck's head looking to the left. In fact, you may be so taken with the picture of the duck that you completely fail to spot the cute rabbit looking to the right. Fake psychics work in a similar way. People often want to believe in the reality of psychic powers, perhaps because they inject a sense of mystery into an otherwise dull world, show that science does not have all of the answers, suggest that human consciousness is a force to be reckoned with, or offer the potential of serious problems being solved with the wave of a magic wand.

In the early 1980s, psychologists Barry Singer and Victor Benassi from the California State University conducted a now classic experiment that demonstrated the power of this principle.[6] Singer and Benassi asked a young magician named

Craig to don a purple robe, some sandals and a 'gaudy' medallion, and then perform magic tricks to groups of students. Some of the time the psychologists introduced Craig as a magician and other times they said that he claimed to possess genuine psychic abilities. Either way, Craig simply performed a series of standard magic tricks that involved him apparently reading people's minds and bending metal. After his performance all of the students were asked whether they thought Craig possessed psychic abilities. A massive 77 per cent of those in the 'Craig is a psychic' group thought that they had seen a display of genuine paranormal phenomena. But more amazingly, 65 per cent of those in the 'Craig is a magician' group also thought he was psychic. It seems that when people are making up their minds about how to perceive the impossible, a purple robe, some sandals and a medallion go a long way.

In the same way that a deep love of ducks can drive people to completely miss the rabbit, so a strong need to believe in psychic powers can cause some people to watch individuals like Hydrick and be totally blind to the possibility of trickery.

Hydrick did all sorts of things to sell the world a duck. He evoked images of the mysterious East by wearing martial arts gear, occasionally called himself 'Song Chai', and made up stories about his encounters with Master Wu. Had he put on a top hat, announced himself as 'Magic Jimbo' and spoken about spending time with David Copperfield, it would all have been very different.

It was also about the type of abilities that he appeared to possess. Early in his career Hydrick experimented with different types of demonstrations. At one point he would apparently cut a piece of string in half, place the ends inside his mouth, claim to be rearranging the atoms and then show that the two pieces had magically joined back together. When he performed the demonstration people (quite correctly) thought that it looked like a magic trick and so it was quickly dropped from Hydrick's repertoire. Cutting and restoring a piece of string set off mental 'this is a magic trick' sirens and encouraged people to go in search of the rabbit. In contrast, moving pencils with mind power matches people's preconceptions about the paranormal and so encourages them to see the duck.

Hydrick also acted as if his powers were genuine. Most of the people who believe in the reality of psychokinesis think that such abilities are both energy sapping and elusive. Hydrick exploited these ideas by often acting as if the demonstrations were a drain on his mental resources, taking a long time before making a page turn or a pencil move, and sometimes even failing completely. He could have easily moved the objects without the slightest strain whenever he wanted but that would have looked like a magic trick.

Finally, he often appeared to bring out people's latent psychic ability by having them believe that their mental powers were responsible for moving the pencil. This is a common ploy used by fake psychics because it has enormous emotional appeal. Many people want to believe that they do indeed have incredible powers and so when they appear to encounter proof of this ego-enhancing concept they become especially reluctant to look behind the curtain and find out what is really going on.

Hydrick walked like a duck and sounded like a duck. Because of this, lots of people assumed that he was the real deal and didn't even consider the possibility of quackery.

Although some of the people watching him didn't even think about fakery, many others would have been far more sceptical. Perhaps they didn't believe in psychic ability, or did believe but were sceptical about Hydrick's particular claim. Whatever their point of view, Hydrick fooled some of these people using a second principle.

Take the Road Less Travelled

Time for two quick puzzles. Here is the first one. Can you add just one line to the following statement in order to make it correct?

$$I0 \ I0 \ II = I0.50$$

Now for the second puzzle. The illustration below shows the number nine represented as a Roman numeral. Can you convert this into the number six by just adding a single line?

IX

You probably assumed that the answer to the first puzzle required some clever mathematical thinking, and that the solution to the second involved Roman numerals. The puzzles are specifically designed to make you think like that. In fact, the solution to the first puzzle involves time, not mathematics. To make the statement correct, all you have to do is add a short line over the second 'I', thus converting the number 'I0' into the word 'TO':

$$I0 \ TO \ II = I0.50$$

Now the equation reads 'ten to eleven is the same as ten fifty'. To solve the second puzzle you draw an 'S' in front of the IX to transform it into the word 'SIX'.

Many people struggle with these types of puzzles because they require lateral thinking. The same principle prevents them from figuring out how Hydrick performed his miracles. Ask people how they would go about making a pencil mysteriously move and they will come up with various ideas. They might, for example, suggest tying a thin thread to it. Or they might think of putting a metal bar inside it and moving a magnet under the table. Or, they might even suggest experimenting with static electricity. However, people just don't tend to think about secretly blowing on the pencil. In the same way that most people struggle with the puzzles above because they don't think about the equation being about time, or that a line in the shape of the letter 'S' would make the word SIX, so Hydrick fooled some sceptics by using a method that didn't cross their non-lateral minds.

Of course, this principle is not going to fool everyone. After all, some people are naturals when it comes to thinking outside the box, while others know a thing or two about trickery, and so would have considered the 'blowing' option. To crack these tougher nuts, Hydrick needed to employ the next principle.

Cover Your Tracks

Watching film of Hydrick in action is fascinating, and reveals just how skilled he was. He uses two main techniques to discourage the 'doesn't he just blow on it?' brigade. First, Hydrick had spent months learning how to carefully control his breath, allowing him to produce perfectly timed puffs of air that took a few moments to reach the objects. The slight time delay between puff and impact gave him time to turn his head around, thus ensuring that he was looking away from the object when it moved. Second, he didn't blow directly at the objects, but rather at the surface of the table. The air currents were then travelling along the tabletop, hitting the objects and causing them to move. This technique ensured that there was never a direct path between Hydrick's mouth and the object. Together, these techniques were extremely deceptive, and allowed him to cover his tracks and encourage those considering the 'blowing' hypothesis to jettison the idea.

When he appeared on *That's Incredible!* Hydrick encountered the toughest type of spectator – the informed sceptic. Host John Davidson was suspicious that Hydrick might be cheating, had figured out that he was breathing on the objects, and did not have the wool pulled over his eyes by Hydrick's head turn and tabletop blowing. To fool Davidson, Hydrick used the fourth and especially deceptive technique.

Change the Route

Our brains are very poor at coping with problems in which the correct answer changes from one moment to the next, and instead like to think that there is a 'one size fits all' solution. Fake psychics like Hydrick exploit this assumption by switching methods when they repeat a demonstration. If one performance rules out one method, and a second performance rules out a second method, spectators assume that neither method accounts for either performance and so conclude in favour of a miracle.

Hydrick's performance on *That's Incredible!* is a classic demonstration of changing the route. When Davidson expressed his scepticism, Hydrick invited the host to place his hand over Hydrick's mouth and yet the pencil still rotated. Why? Because Hydrick made a quick karate chop in the air and the resulting currents caused the pencil to move. He changed the route, and both Davidson and the viewers were completely fooled.

Hydrick fooled different people for different reasons. Some believed that he was psychic and so the thought of trickery never entered their duck-loving minds. Others considered the possibility that they were watching a trick, but didn't think of the correct method. Some thought of the correct method but Hydrick's head-turning and indirect blowing made them think that they were wrong. A minority thought of the correct solution and were not fooled by his skilled performance, but were baffled when he switched methods during repeated performances. Still, although highly effective, all of these principles would have a high chance of failure if it weren't for the fifth, and most important, factor. But for now, a fun trick . . .

THE PSYCHOLOGY OF SPOON-BENDING

It's time to apply some of the principles of deception discussed so far to fool your friends and family. Want to
appear to bend a spoon with the power of your mind? Try
following this . . .

1. When you are out in a restaurant or round at a friend's
house for dinner, secretly remove one of the spoons from
the table, put it in your pocket, and go to the toilet.

2. Once hidden away, carefully bend the bowl of the
spoon towards the stem and then bend it back again.
Repeat this process a few times. Two things will begin to
happen. First, the metal around the bend will start to
become very hot – be careful about burning your fingers.
Second, the spoon will eventually develop a very fine fracture line at the point of the bend. As soon as you see the
line, stop bending because even the smallest of movements
will cause the spoon to break in half. You have now created what fake psychics refer to as a 'pre-stressed' spoon.

3. Place the pre-stressed spoon back into your pocket and
return to the table.

4. When people are engaged in lively conversation,
secretly take the spoon from your pocket and place it on
your lap. Then, when the group is engaged in even more
lively conversation, pick up the spoon from your lap and
secretly place it back on the table.

5. When the conversation has ceased being lively, bring up the topic of psychokinesis and claim that as a child you could bend metal with the power of your mind. Explain that you haven't tried it for years, but are prepared to give it a go. If no one is interested, get your coat and go hang out with a more interesting group of people.

6. Assuming there is some interest in your lies, pick up the pre-stressed spoon and place the finger and thumb of your right hand either side of the fracture. By giving the spoon a slight jiggle with your left hand you will find that it easily breaks in half. Hold the two halves together between your thumb and finger as if the spoon were still solid. Then slowly relax your grip and cause the spoon to appear to bend before finally breaking in two.

7. Allow the two halves of the spoon to drop onto the table with a dramatic clatter. If you are at a friend's house, now is a good time to ask them if the cutlery is especially expensive or has sentimental value. Either way, you now have two options. You can explain how you performed the trick and get your friends to try it with the remaining cutlery. Alternatively, you can claim that it was a miracle, explain that you are thinking of starting a cult, and ask people if they are interested in joining.

This trick is especially effective because people assume that the performance begins when you announce that you are about to bend a spoon with the power of your mind. In reality, it began when you secretly picked up the spoon and placed the fracture in it. This technique, referred to by magicians as 'time misdirection', accounts for the success of many illusions and demonstrations of alleged psycho-kinesis.

People often underestimate the effort that some magicians and fake psychics put in before the start of a performance. For example, British magician David Berglas was once invited to stage a private performance in the third floor apartment of a wealthy London banker. During the performance, Berglas borrowed an empty milk bottle from the banker, attached it to a long piece of string, and carefully lowered it out of the apartment window. Next, Berglas picked up a pear from the fruit bowl, and apparently made it vanish into thin air. The banker was then asked to carefully retrieve the bottle by pulling on the string, and was amazed to discover that the pear was now inside the bottle, even though it was too large to fit through the glass neck of the bottle. This seemingly spontaneous piece of trickery involved a huge amount of planning. Months before, Berglas had found a pear tree with budding fruits, and placed one of the stems into an empty milk bottle. Over time, the pear grew inside the bottle, giving Berglas his impossible looking object. During the trick, he simply had an assistant stand on the street and switch the bottle that was lowered from the apartment for the duplicate containing the pear, and thus fooled his guests, who assumed that the trick had begun just a few moments before.

In the Land of the Blind

Before progressing to the fifth and final principle of psychic deception, it is first important to turn back the hands of time and find out about one of the most controversial experiments in the history of supernatural science.

In 1890 Mr S. J. Davey announced that he had acquired the gift of mediumship and invited small groups of people to his London home to witness his remarkable abilities. Each group gathered in Davey's dining room and was asked to sit around a table. He then lowered the gaslight and joined the group.

Some of the guests had been asked to bring along some school slates, and at the start of the séance Davey placed a piece of chalk on one of them and put the slate under one corner of the table, with the edges protruding. He then held one edge and invited a member of the group to grasp the opposite side. Pushing the slate tightly against the underside of the table, Davey asked the spirits, 'Will you do anything for us?' Within a few moments some mysterious scratching sounds were heard, and when the slate was withdrawn the word 'Yes' was clearly written across its surface.

Encouraged by his success, Davey moved onto the second part of the séance. After the group had searched the room for any evidence of trickery, he extinguished the gaslights, and asked everyone to hold hands and join him in summoning the spirits. Slowly, a pale blue light materialized over Davey's head. The light then developed into a full-form apparition that one guest later described as 'frightful in its ugliness'. After this spirit had faded into the darkness, a second streak of light appeared and slowly developed into 'a bearded man of Oriental appearance'. This new spirit bowed and moved just a few feet from those present, its complexion was 'not dusky, but very white; the expression was vacant and listless'. The spirit then floated high into the air, and vanished through the ceiling.

Night after night people left Davey's house convinced that they had made contact with the spirit world. In reality, Davey did not possess the ability to summon the spirits. Instead, he was a conjuror who had used his magical expertise to fake all of the phenomena. However, unlike almost all of the other fake mediums of his day, Davey was not interested in fame or fortune. Instead, his guests had been unsuspecting participants in an elaborate and cleverly conceived experiment.

In Davey's day, many mediums claimed to be able to make the deceased write on school slates and materialize in front of people's eyes. Those attending these demonstrations frequently found them convincing and came away certain that the soul survived bodily death. Davey was deeply sceptical and believed that the public were being fooled and fleeced by unscrupulous con artists. There was, however, one small problem. Many of the people attending the séances described witnessing incredible phenomena that couldn't have been caused by trickery. Davey decided to conduct his own fake séances to discover what was going on.

In the same way that Korem learned to replicate Hydrick's tricks, so Davey schooled himself in the devious ways of fake mediumship. Night after night Davey performed for his unsuspecting guinea pigs, and then asked each of them to send him a written account of the evening. He asked them to make their testimony as complete as possible and describe everything that they could remember. He was stunned to discover that people frequently forgot or misremembered information that was central to his trickery.

The slate-writing demonstration is a good example. Before the séance Davey attached a small piece of chalk to a thimble and slipped it into his pocket. When one of his guests took out a slate Davey slipped the thimble onto his finger. Then, when the slate was held beneath the table Davey wrote the word 'yes' on the underside. He then removed the slate and, by showing the upper face only, confirmed that there was no message. As the slate was replaced under the table, Davey turned the slate over, ensuring that the writing was now pushed against the underside of the table. When it was removed a second time, the word 'Yes' had mysteriously appeared. When participants later described the demon-stration, the all-important removal and replacement of the slate vanished from their memories, with the guests firmly believing that the slate was placed under the table and had remained there until the spirit writing appeared.

There were also the alleged materializations. Before the guests arrived Davey hid a large amount of fake spirit apparatus in one of his dining room cupboards. Before extinguishing the gaslight, he invited the group to thoroughly search the séance room. When he saw someone about to look in the cupboard containing his spirit stash, he quickly diverted their attention by inviting them to search him for any hidden paraphernalia. When the room was plunged into darkness, Davey's trusted friend, Mr Munro, quietly sneaked into the room, retrieved the objects hidden in the cupboard, and used

them to fake various spirit forms. The 'apparition of frightful ugliness' was a mask draped in muslin and treated with
luminous paint, while the 'bearded Oriental' was the result
of Munro dressing up ('a turban was fixed upon my head, a
theatrical beard covered my chin, muslin drapery hung about
my shoulders') and illuminating his face with a weak phos-
phorescent light. Munro later noted that though 'the pallor of
my face was due to flour, "the vacant and listless expression"
is natural to me'. To create the illusion that the spirit levitated
and then vanished, Munro stood on the back of Davey's chair,
lifted the light high above his head, and extinguished it when
he reached the ceiling. In the same way that people misremembered the slate writing, so they were convinced that they had
thoroughly searched Davey's dining room and completely
forgot that they had not looked inside one of the cupboards.

In 1887 Davey published a 110-page dossier cataloguing
a huge number of these errors, and concluded that people's
memories for apparently impossible events cannot be trusted.
The report caused a sensation.[7] Many leading Spiritualists,
including the co-creator of the theory of evolution Alfred
Russel Wallace, refused to believe Davey's findings.[8] Desperate to find out how all of his tricks were performed, Wallace
declared that unless all of the fakery was explained, he would
be forced to conclude that Davey possessed genuine mediumistic powers and was deceiving the public by instead claiming
to be a magician. Davey contracted typhoid fever and died in
December 1890, aged just 27. Soon after his death, Munro
and others explained how they had faked all of the phenomena, but Wallace still didn't accept it.[9] In a long article he
presented detailed descriptions of other séances that he had
attended wherein such trickery would have been impossible.

Davey's supporters noted that there was no reason to believe that Wallace's testimony was any more accurate than the ones produced by those attending Davey's fake séances.

Air-brushing the Past

Davey's findings are an astonishing example of the fifth and final principle used by Hydrick and other fake psychics to fool the world. Many people think that human observation and memory work like a video recorder or film camera. Nothing could be further from the truth. Take a look at the picture below of two people sitting at a table.[10]

In a moment, I would like you to turn over the page and look at a second picture. Although the new picture will appear very similar to the one below, a large part of the image has been altered. Try to spot the change. To make things as fair as possible, feel free to flick between the two pictures. OK, away you go.

Most people struggle to identify the difference, even though it is staring them in the face. If you haven't spotted it, let me put you out of your misery. In the second picture the bar at the back of the photograph is much lower. Don't feel upset if you didn't spot the change. In fact, the vast majority of people struggle to see it. Psychologists refer to this rather curious phenomenon as 'change blindness' and the effect is a direct result of the way in which your visual processing system works.

When you first saw the picture you probably had the feeling that you were seeing all of it in a single instant. This is a compelling illusion generated by your brain. In reality, the ability to form such an instantaneous perception would take a massive amount of brainpower. Rather than evolve a head the size of a planet, your brain uses a simple shortcut to create the feeling of instantaneous perception. At any one moment, your eyes and brain only have the processing power to look at a very small part of your surroundings. To make up for this somewhat myopic view of the world, your eyes unconsciously

dart from one place to another, rapidly building up a fuller picture of what is in front of you. In addition, to help ensure that precious time and energy aren't wasted on trivial details, your brain quickly identifies what it considers to be the most significant aspects of your surroundings, and focuses almost all of its attention on these elements.

Conceptually, it is as if you are standing in a darkened sweet shop with a torch, and getting a rough idea of what sweets are on the shelves by quickly moving the beam from one spot to another, and then concentrating on the jars that hold your favourite kinds of confectionery. However, rather than letting you know that you are not looking at the entirety of your surroundings in an instant, your brain pieces together an image based on its initial scan of the area and presents you with the comfortable feeling of constantly being aware of what is going on around you.

In the case of the picture, eye tracking studies show that the bar receives scant attention, with most people focusing on the faces of the two people (with around 55 per cent of people wondering what on earth the woman sees in the guy). However, despite this selective looking, your visual system provides you with the impression that you are constantly seeing the entire picture, thus explaining why you could not spot the difference.

This process is taking place every moment of your waking life. Your brain is constantly choosing what it believes to be the most significant aspects of your surroundings and paying very little attention to everything else. By making important actions seem unimportant, fake psychics are able to use this principle to make key aspects of their performance vanish from spectators' minds. For example, when Davey first withdrew the slate from under the table he seemed to be checking for a spirit message. Because of this the movement of the slate seemed unimportant and so was quickly forgotten by his guests. Similarly, when performing his stunts, Hydrick would briefly glance at the objects, secretly blow and then look away. Because the glance seemed so trivial, people would forget about it and later be convinced that Hydrick looked away from the objects throughout his demonstrations.

The first four principles of psychic deception – selling the duck, taking the road less travelled, covering your tracks, and changing the route – ensure that people do not figure out the solution to the tricks that are happening right in front of their eyes. The fifth principle – air-brushing the past – ensures that they are unable to accurately remember what happened. Without spectators realizing it, important details vanish from their minds and they are then left with no rational way of explaining what they have witnessed.

THE GURU AND THE REFRIGERATOR

A few years ago a colleague and I travelled to India to investigate top Godman Swami Premananda.[11] Born in 1951, Premananda claims that his religious calling became apparent when he was a teenager and a saffron-coloured robe suddenly materialized on his body. Since then, Premananda has performed his alleged miracles on an almost daily basis, materializing objects in his bare hands and regularly regurgitating egg-shaped stones. In the early 1980s Premananda established a religious retreat in a remote part of Southern India, and at the time of our visit this self-contained village was home to the guru and about 50 of his followers. Drawn from across the world, this merry band of devoted disciples were convinced that their leader's miracles were genuine and had dedicated their lives to his teachings.

My initial glimpse of Premananda was somewhat strange. On the first day of our visit I went to the retreat shop to buy a cold drink. The owner said that unfortunately his refrigerator had broken and that he was waiting for Premananda to solve the problem. I instantly conjured up a mental image of Premananda's followers cramped into a meeting hall with their guru leading the group in refrigerator-based prayer. A few moments later the shop door swung open and in walked Premananda clutching a bag of tools. The Swami yanked the refrigerator away from the wall, took a spanner out of his bag, and started tinkering away at the back of the machine. Within minutes the refrigerator burst into life. Sensing that his work here was done, Premananda quickly re-packed his tools, bought a chocolate bar, and left.

That afternoon we were informed that Premananda would meet us at six o'clock the following morning to demonstrate his paranormal powers. Early the next morning I dragged myself off the wooden plank that constituted my bed and made my way to the meeting hall. Six o'clock came and went. As did seven o'clock, followed by eight o'clock. It seemed that Premananda was playing the 'guru game'; testing our level of devotion by arriving several hours after an agreed time. (When I play the same game with my students it is referred to as 'unprofessional behaviour'.) After four hours waiting in an increasingly hot and sticky hall, I decided that enough was enough and made my way towards the exit. As if by magic, the door swung open and in walked Premananda, surrounded by a small group of followers.

The Godman smiled and quickly made a sweeping motion with his hand. A small stream of 'vibhuti' – a fine ash used in Hindu worship – started to trickle from his fingertips. A few moments later the ash ceased and Premananda appeared to pluck two small gold trinkets from thin air. Miracles over, I handed my Polaroid camera to one of the devotees and suggested that we all step outside for a group photograph. The resulting image clearly showed an odd purple haze surrounding the group and two additional blobs of purple directly above Premananda and me. Premananda looked at the photograph and modestly pointed out that many religions associated the shade of purple with sainthood.

Careful observation of the guru at work suggested that he had hidden the objects that he miraculously found in the folds of his garment, and was secretly picking them up when people weren't watching. When we eliminated the possibility of this by placing a clear plastic bag around his hand, the materializations suddenly dried up.

And what about the purple haze on the photograph of Premananda? When I returned to Britain I took the photograph to the Polaroid laboratories. The technician explained that when a Polaroid photograph is ejected from the camera, pouches containing developing chemicals are broken and the chemicals are dragged across the image. The technician then looked at the code number on the back of my photograph, consulted a big book of numbers, and revealed that the chemicals would have been past their sell-by date and therefore prone to a purplish discolouration. As a result, the scientific community has been reluctant to view the image as compelling evidence of sainthood. Personally, I am more convinced.

Field footage of the Premananda test
www.richardwiseman.com/paranormality/Premenanda.html

Davey's ground-breaking work constitutes the very first experiment into the reliability of eyewitness testimony. Since then, psychologists have carried out hundreds of such studies that have demonstrated that the same type of selective memory clouds our ability to recall everyday events.

Around the turn of the last century, German criminologist Professor von Lizst conducted some dramatic studies on the subject.[12] One such study was staged during one of von Lizst's lectures and began with him discussing a book on criminology. One of the students (actually a stooge) suddenly shouted out and insisted that von Lizst explore the book from 'the standpoint of Christian morality'. A second student (another stooge) objected and a fierce argument ensued. The situation went from bad to worse: the two stooges started to trade punches and eventually, one of them pulled out a revolver. Professor von Lizst tried to grab the weapon and a shot rang out. One of the students then fell to the ground and lay motionless on the floor.

Professor von Lizst called a halt to the proceedings, explained that the whole thing was a set-up, had his two stooges take a bow, and quizzed everyone about the event. Von Lizst was amazed to discover that many of his students had become fixated on the gun (a phenomenon that psychologists now refer to as 'weapon focus') and so, without realizing it, had forgotten much of what had happened just a few moments before, including who started the argument and the clothing that the protagonists were wearing.

In the 1970s psychologist Rob Buckhout conducted a similar experiment, staging mock assaults in front of over 150 witnesses.[13] Again, the witnesses tended to focus on what they thought was important – the nature of the assault – and so failed to remember much other information about the incident. When they were later shown six photographs and asked to identify the perpetrator, almost two-thirds of them failed to do so. On another occasion an American television programme broadcast footage of a mock purse-snatching incident and then asked viewers to try to identify the thief from a six-person lineup. Over 2,000 people called the programme and registered their decision. Even though the footage clearly showed the face of the assailant, just over 1,800 of the viewers identified the wrong person.[14]

A large amount of research has revealed the same finding time and again. We all like to think that we are reliable eye-witnesses. However, the truth of the matter is that, without realizing it, we tend to misremember what has happened right in front of our eyes and frequently omit the most important details.

Your brain is constantly making assumptions about which parts of your surroundings are most deserving of attention and the best way of perceiving what is there. Most of the time these assumptions are correct, and so you are able to accurately perceive the world in a highly efficient and effective way. However, once in a while you will encounter something that trips up this finely honed system. In the same way that a good optical illusion completely fools your eyes, so those claiming psychokinetic abilities perform the simplest of magic tricks but fool you into thinking that you have witnessed a miracle. They subtly discourage you from considering the possibility of deception, use sneaky methods that you would never consider, and ensure that any possible evidence of trickery is quickly airbrushed out of your memory. Seen in this way, rotating pencils and bending spoons are not proof of the impossible, but are instead vivid reminders of just how sophisticated your eyes and brain really are. The people performing these demonstrations do indeed have remarkable powers, but their skills are psychological, not supernatural.

4. TALKING WITH THE DEAD

In which we meet two young girls who created a
new religion, discover what happened when the world's
greatest scientist confronted the Devil, learn how to
commune with non-existent spirits and unleash
the power of our unconscious minds.

It is 10 p.m. and we are just about to start the session. Ten unsuspecting members of the public and I are sitting around a wooden table in the front room of a house in London's East End. The room is in near darkness, illuminated only by a couple of candles on the mantelpiece. I ask everyone to lean forward and place their fingertips lightly on the tabletop, take a deep breath and call upon the spirits to join us. Nothing happens. I tell everyone not to become dispirited and to suspend any scepticism that they might have. Once again I speak into the darkness and ask the spirits to make their presence known by moving the table. After a short time the table gives a small, but real, shudder. It is a good sign, and I have a hunch that we are all in for an interesting night.

Over the course of the next 30 minutes the table shudders several more times. A man in the group then says that he is going to have to nip to the toilet. As he stands up, the tabletop emits a tremendous creaking sound and suddenly tips up on two legs. It is a dramatic movement, and it feels as if someone has kicked the table from below. Several people in the group scream and the man decides that perhaps his trip to the toilet isn't that important after all. All four legs of the table return to the ground and the table starts to skid from one side of the room to the other, sometimes pinning members of the group to the wall. After about an hour the movements suddenly cease and we solemnly thank the spirits for making their presence known. The candles are blown out, the lights are turned on, everyone discusses the strange events that they have just experienced, and the man finally gets to go to the toilet.

I have staged many such séances over the years and the results are always the same. Regardless of whether the group consists of believers or sceptics, the table always moves. Even if everyone takes turns removing their fingers from the table-top, the table continues to tip and shake.

Table-tipping was first practised in Victorian parlours throughout Britain, and the phenomenon is as puzzling to the modern-day mind as it was to those living then. But when it comes to talking with the dead, table-tipping is just the tip of the iceberg. In other types of séance, the Victorians asked the deceased to spell out messages by moving an upturned glass towards alphabet cards and even to scribble words directly onto pieces of paper. Investigations into these curious phenomena yielded surprising insights into the power of the unconscious mind, the fundamental nature of free will, and how to be a better golfer.

This remarkable story starts with two sisters who managed to fool the world.

Clever Like a Fox

Around the turn of the last century Thomas Hardy wrote a poem in which he described witnessing God's funeral. Hardy's verses vividly express the sadness experienced by the religious if they come to doubt the existence of a divine creator.

Throughout the nineteenth century more and more people came to experience the painful feelings described by Hardy as established religion came under a serious and sustained attack. The great Scottish thinker David Hume set the ball rolling by criticizing the then sacrosanct idea that alleged evidence of design in nature constituted compelling proof of God, with Hume eventually publishing his ideas in a blasphemous book entitled *Dialogues Concerning Natural Religion*, originally considered so controversial that it was published anonymously and didn't even carry the publisher's name. Hot on his heels was the English philosopher John Stuart Mill, who argued that the public were a fairly rational bunch, and so should be allowed to choose their religious beliefs, or not, without any interference from the state. And then along came Charles Darwin with his dangerous idea that men and beasts may not be quite so different after all.

Organized religion began to feel the pinch. For centuries priests and clergymen had fought the Devil, but now found themselves facing a new and far more daunting enemy – congregations that dared to demand evidence for their God.

They proved a tough crowd. The Victorians were enjoying the benefits of unprecedented scientific advances, from steam engines to sewing machines, photography to petrol, telephones to tarmac, phonographs to paperclips, and jelly babies to ice cream. Suddenly, age-old stories about a man who could feed 5,000 people with just five loaves of bread and two small fish simply failed to cut the mustard. To many it seemed that the church had little to offer but blind faith and somewhere warm to sit on Sundays.

As religion rapidly lost ground to rationality, the end-game seemed inevitable. Indeed, some writers were happy to declare the battle already over, with perhaps the most unequivocal statement coming from German philosopher Friedrich Nietzsche: 'God is dead. God remains dead. And we have killed him.' Predictably, believers were somewhat more optimistic. Although well aware that their creator was on the critical list they hoped that, to paraphrase Mark Twain, the reports of his death were greatly exaggerated.

Feeling increasingly under attack, the religious did what they had always done in difficult times. They put their heads down, placed their hands together and prayed for a miracle. On 31 March 1848 God appeared to answer their prayers.

Hydesville is an unassuming hamlet about 20 miles east of Rochester, New York.[1] In December 1847, John and Margaret Fox moved into a small house on the edge of the hamlet with their two daughters, 11-year-old Kate and 14-year-old Margaretta. Within a few months the Fox family life was disturbed by a series of odd events. Bedsteads and chairs started to shake, ghostly footsteps were heard moving through the house, and on occasion the entire floor of the property vibrated like a giant drum skin. After John and Margaret's investigations failed to provide an explanation for these apparently supernatural happenings, they found themselves forced to conclude that their new home was haunted by an 'unhappy restless spirit'.

On 31 March 1848 the family had gone to bed early in an attempt to get a good night's rest, without any ghostly shenanigans. Unfortunately, it was not to be. Within a few moments of them settling down, the disturbances started. Rather than simply enduring another night of endless shaking and knocking, young Kate decided to attempt to communicate with the spirit. Making the rather pessimistic assumption that their unwelcome guest might be the Devil himself, Kate spoke into the darkness and asked 'Mr Splitfoot', as she'd decided to name him, to copy her actions. She clapped her hands three times. A few seconds later three raps mysteriously emanated from the walls of the house. Contact had been made. Intrigued, Margaret Fox then nervously asked the entity to rap out the ages of her children. 11 knocks were heard for Kate. Pause. Then 14 knocks for Margaretta. Pause. Then three knocks. Three knocks? The entity was well informed – Margaret had had a third child who had died several years before, aged three.

The spiritual chit-chat continued long into the night, with the family eventually developing the now infamous 'one rap for yes, two raps for no' code, and then using it to establish that the entity was a 31-year-old man who had been murdered in the house a few years before their arrival, and whose remains were currently buried in their cellar. The following night, John Fox attempted to dig up the cellar floor in search of bones, but was forced to abandon the work when he reached the water level.

Word of the strange happenings quickly spread to surrounding towns, resulting in hundreds of people coming to Hydesville to experience the raps for themselves. Many of them got to communicate with the spirit, which only served to further feed the ghostly gossip now rapidly moving across New York. Within a few months the constant stream of visitors and rapping took its toll, with Margaret Fox's hair turning white through worry and her husband being unable to work. Eventually they decided that it was in everyone's best interests to move their children away from the spirit-infested house. Kate was sent to nearby Auburn and Margaretta to Rochester. But the seeds had already been sown that would change the course of history.

The various spirits conjured up by Kate and Margaretta followed the two young girls, with rapping breaking out in their new locations. In Rochester, a long-standing family friend and committed Quaker named Isaac Post had an idea. The rapping code was proving a rather time-consuming, and sometimes confusing, way of eliciting information from the spirit world. Would it be possible, Isaac wondered, to create a more accurate type of communication? One evening he invited Margaretta to his house and asked whether she would mind experimenting with a new system. He drew the letters of the alphabet on pieces of paper, and explained to the spirits that he would ask a question and then point to each piece of paper in turn. To communicate whatever was going through their discarnate mind, the spirits simply needed to rap when he was pointing to an appropriate letter. Isaac's instant messaging with the dead proved a hit and soon resulted in the first fully-formed communication from beyond the grave. Not one for small talk, the spirits issued a firm and frank directive:

Dear Friends, you must proclaim this truth to the
world. This is the dawning of a new era. You must not
try to conceal it any longer. When you do your duty
God will protect you and good spirits will watch over
you.

Convinced of the genuineness of the messages, Isaac enthu-
siastically embraced the new religion of 'Spiritualism' and set
about converting his fellow Quakers.

From a psychological perspective, the creation of
Spiritualism was a stroke of genius. Whereas the established
churches had tried to combat the rise in rationality by stress-
ing the importance of faith, Spiritualism changed the very
nature of religion. In an age that was obsessed with science
and technology, Spiritualism not only offered proof of an
afterlife but, on a good night, allowed people to apparently
communicate with their deceased loved ones.[2] Other religions promised the tantalizing possibility of life after death.
Spiritualism delivered the goods. This combination of rational
and emotional appeal proved overwhelming and within just a
few months the new religion was sweeping across America.

The Fox sisters quickly gained celebrity status and received
invitations to demonstrate their amazing mediumistic abilities
in public shows and private gatherings. They chatted with the
spirits about any topic put to them, with newspaper reports
describing how one moment they were being consulted on the
weightiest of philosophical and religious issues while the next
they were discussing railway stocks and love affairs.

From the very start, Spiritualism shared many of the central tenets of Quakerism, including support for the abolition
of slavery, the temperance movement and women's rights.

The new religion also adopted the Quakers' non-hierarchical structure. Out went the idea of high priests and untouchable clergymen, and in came the notion of spiritual democracy, with followers being encouraged to gather together and experiment with different ways of talking to the dead. And gather they did. In parlours across America and Europe small groups of Spiritualists would meet up and try to make contact with their deceased loved ones (or indeed any other spirit who might be kind enough to drop in).

When it proved difficult to replicate the raps produced in the presence of the Fox sisters, the groups started to experiment with more reliable forms of communication. By far the most popular technique to emerge was that of table-turning. In a typical session, people would sit around a small table, place their fingertips lightly on its surface, turn down the gaslight, sing a few hymns, and start to summon the spirits. After a while everyone would start to feel the wooden tabletop creak and shiver beneath their hands. A little more hymn singing and the table would suddenly start to tip and move, as if being pushed and pulled by spirits. According to reports from the time, on a good night the table appeared possessed, dancing around the room, climbing affectionately onto people's laps, and sometimes even aggressively pinning them up against the wall. Table-turning spread like an epidemic and soon hundreds of thousands of people were passing their evenings transforming a common piece of household furniture into a conduit to the afterlife.

'I was the First in the Field and I Have a Right to Expose it'

With the rapid growth in the number of mediums, the pressure of trying to make ends meet in an increasingly crowded market place eventually took its toll on Kate and Margaretta Fox. The two of them gradually formed a somewhat different kind of bond with the spirit world and by the late 1880s both were drinking heavily. In October 1888 they decided that enough was enough and travelled to New York City to make a dramatic announcement.

Selling her story to the *New York World* for an alleged $1,500, Margaretta came clean and confessed that the two of them had faked the entire affair.[3] A new convert to the Catholic Church, she could take the guilt no longer. According to her, the strange noises initially experienced at Hydesville were actually due to nothing more than an apple, a piece of string and a naive belief in the honesty of children:

> When we went to bed at night we used to tie an apple to a string and move the string up and down, causing the apple to bump on the floor, or we would drop the apple on the floor, making a strange noise every time it would rebound. Mother listened to this for a time. She would not understand it and did not suspect us as being capable of a trick because we were so young.

Margaretta went on to explain that the 'apple on a string' technique was only effective in darkness and so the sisters quickly devised a different way of creating raps in daylight:

> The rappings are simply the result of a perfect control of the muscles of the leg below the knee, which govern the tendons of the foot and allow action of the toe and ankle bones that is not commonly known . . . With control of the muscles of the foot, the toes may be brought down to the floor without any movement that is perceptible to the eye. The whole foot, in fact, can be made to give rappings by the use only of the muscles below the knee. This, then, is the simple explanation of the whole method of the knocks and raps.

After reflecting on the stress that she had endured as a result of a life of deception, Margaretta provided an unequivocal statement about the nature of the new religion that she had helped create:

> Spiritualism is a fraud of the worst description . . . I want to see the day when it is entirely done away with. After my sister Katie and I expose it I hope Spiritualism will be given a death blow.

Later that week Margaretta silenced those Spiritualists who had been sceptical about her confession by appearing before a packed auditorium at the New York Academy of Music and demonstrating her remarkable ability to produce raps at will. Did her dramatic confession have the desired effect? Did the estimated eight million Spiritualists in America alone throw up their hands in horror and desert their new-found faith? Sadly, the only real impact of the confession was to distance the sisters from their supporters. The vast majority of Spiritualists were eager to cling to the comforting thought that they might survive bodily death, and they were not going to let a couple of rambling alcoholics stand in the way of immortality. But although Margaretta tried to retract her remarks shortly after confessing all, for the Fox sisters at least, the damage had been done. Increasingly distanced from the movement that they helped to create, both sisters died in poverty a few years later and were buried in pauper's graves. Neither made contact from the spirit world.

By now, the genie was out of the bottle. Tables were turning all across America and Britain. Even more impressively, some of them were starting to actually talk.

Interview with historian Peter Lamont
www.richardwiseman.com/paranormality/PeterLamont.html

The Devil's Mouthpiece

The idea was simple enough. If a table could be moved by spiritual energy, surely it could also be used as a way of actually getting a message from the other side? Initially people started asking questions during table-turning sessions and employing a variant on the Fox sisters' code to interrogate the spirits – one tip for yes and two for no. When this proved somewhat time-consuming, people followed in the footsteps of Isaac Post, calling out the letters of the alphabet and asking the spirits to spell their message by tipping the table at appropriate points. Accounts suggest that these sessions could be highly emotionally charged affairs, as the following description from Edinburgh in 1871 shows:

> At a particular stage of the proceedings the table began to make strange undulatory movements, and gave out a curious accompaniment of creaking sounds. Presently my friend remarked that the movement and sound together reminded him of a ship in distress, with its timbers straining in a heavy sea. This conclusion being come to, the table proceeded to rap out: 'It is David.' Instantly a lady burst into tears, and cried wildly: 'Oh, that must be my poor, dear brother, David, who was lost at sea some time since'.[4]

Many were far from happy with the idea of talking furniture. Perhaps the most critical voices came from clergymen who became convinced that the Devil was lurking in tables throughout the land. In 1853 the Reverend N. S. Godfrey took it upon himself to prove this by getting information straight from the horse's mouth. Presenting the work in his book, *Table Turning: the Devil's modern masterpiece,* Godfrey described a remarkable episode in which he had a group of table-tippers chat with their four-legged friend, and then asked the table if it contained an evil spirit.[5] The table indicated that it didn't. Realizing that the Devil would be less than straight with his answers, Godfrey asked for the good book to be brought forth. While the table was vibrating, the Bible was placed on its surface, and the moment contact was made with the tabletop the shaking suddenly stopped. Godfrey took this as a sign that the table might be possessed. It would be nice to think that after an hour or so of intense cross-examination the table eventually broke down and admitted all. However, never one to jump to conclusions, Godfrey asked two of his ecclesiastical brethren, the Reverend Gillson and the Reverend Dibdin, to replicate his experiment with different tables. When they obtained the same result Godfrey went public, denounced the phenomenon as the Devil's mouthpiece and warned the public to distance themselves from the potential wooden menace lurking in their parlours and dining rooms.

The rather tiresome procedure of having to list the alphabet and wait for a reply eventually spelled the death of talking tables. Rather than fade into the metaphorical ether, Spiritualism did what it always did. It bowed to market forces and quickly developed a new and improved procedure for talking to the dead. To speed things up, people wrote the letters of the alphabet on small pieces of paper and arranged them in a circle on a table. They would then place their fingertips on an upturned glass and question the spirits. An invisible force would then push the glass from one letter to another as the spirits spelled out their replies. This new method of communication spread like wildfire. quickly resulting in several manufacturers producing commercial versions of the system, referred to as Ouija boards (most probably derived from the French and German words for 'yes'). For a relatively small amount of money people could abandon their scraps of paper and upturned glass for a professionally printed board and little wooden platform on casters (called a 'planchette'). From its introduction in 1891 the Ouija board proved an instant hit and soon formed a mainstay of parlour entertainment across America and Europe.

But as the public began to look for faster ways of chatting with the dead, the need for speed soon overcame even the Ouija board. The front leg of the planchette was eventually replaced by a pencil, and a piece of paper took the place of the Ouija board. People would again place their hands on the planchette, but this time any movement would result in the pencil writing directly on the paper. Suddenly the spirits could dictate messages to the here and now. After further experimentation it was discovered that even this system was an unnecessary burden, and that a small number of people could simply hold a pen or pencil, open their hearts to the spirit world, and receive messages directly from the deceased. This small band of communicators claimed that they were not consciously controlling their own hands, with several writers using this new system to indulge in so-called 'automatic writing', allegedly channelling religious texts, poems and prose from the spirit world.

By the 1920s the world had moved on and belief in Spiritualism fell into decline. The advent of radio and cinema meant that people no longer felt the need to spend their evenings waiting for a message from the dearly departed. This decline continued throughout the twentieth century and nowadays the small number of Spiritualist churches still in operation are usually run by a handful of elderly people who appear only hours away from discovering the reality about life after death.

During the heyday of Spiritualism thousands of people claimed to have contacted the dead via table-turning, Ouija boards, and automatic writing. Did their testimony represent compelling evidence of life after death, or is there a scientific explanation for these apparent spiritual intrusions? A small number of Victorian scientists were eager to examine the curious phenomena and discover what was really going on. Perhaps the most insightful investigation was conducted by a man who is now widely acknowledged to be one of the world's greatest scientists.

Enter Michael Faraday, champion of the invisible.

One Day Sir, You May Tax It

Born in South London in 1791 to a family of modest means, Michael Faraday became fascinated by all things scientific at an early age. His diligence and curiosity soon caught the attention of leading scientist Humphry Davy, resulting in Faraday being given a position at London's prestigious Royal Institution aged just 21.

Faraday worked at the Institution throughout his life, investigating a wide and eclectic range of topics. He invented the world-famous Bunsen burner, discovered that coal dust was the major cause of mining explosions, advised the National Gallery on how best to clean its art collection, and gave a series of popular public lectures on the science of the burning candle ('There is no more open door by which you can enter into the study of natural philosophy than by considering the physical phenomena of a candle').

He is perhaps best known for his ground-breaking investigations into the relationship between the invisible and mysterious forces of electricity and magnetism. After investing hours of bench time tinkering with various apparatus, Faraday's breakthrough came when he bent a piece of wire into a loop, moved a magnet through the centre of it and discovered that the movement of the magnet induced an electric current in the wire. This simple demonstration revealed a fundamental link between electricity and magnetism and paved the way for modern-day electromagnetic theory. Albert Einstein was so impressed with the work that he kept a photograph of Faraday on his study wall as a source of inspiration. Ever the practical man, Faraday immediately set about exploring possible applications for his discovery, eventually creating a forerunner of a modern power generator. When William Gladstone, the British Chancellor of the Exchequer, heard about this newfangled device he quizzed Faraday about the practical value of electricity. Faraday famously replied, 'One day sir, you may tax it.'

Faraday was also serious about his religion, serving as a lay preacher in an obscure offshoot of the Scottish Presbyterian Church known as the Sandemanians. His membership of the church caused him to refuse the presidency of the Royal Society and a knighthood on the grounds that he did not believe Jesus would accept such honours. He also turned down the government's request to develop poison gases for the Crimean War on ethical grounds, and wouldn't buy insurance as he believed it reflected a lack of faith. His religious beliefs may have also played an important role in his discovery of electromagnetism. Believing that one God was responsible for the world, Faraday was convinced that all of nature must be interconnected, including the apparently unlinked forces of electricity and magnetism.[6]

Given Faraday's expertise in harnessing invisible forces and interest in matters spiritual, it is not surprising that he was drawn to table-turning. In 1852 he assembled a group of trustworthy and successful table-movers, and carried out a cunning three-stage plan that still stands as a textbook example of how to investigate the impossible.[7]

In the first stage of his investigation Faraday glued together a bizarre bundle of materials – including sandpaper, glass, moist clay, tinfoil, glue, cardboard, rubber and wood – and secured it to the top of a table. He then asked his participants to put their hands on top of the bundle and summon the spirits. The group had no problem moving the table, meaning that the materials used did not inhibit the work of the spirits. The experiment therefore gave Faraday a free hand to employ the bundle of materials during the second stage of the investigation.

Retreating to his laboratory, he set about constructing several strange bundles. Each consisted of five postcard-sized pieces of cardboard interleaved with small pellets of specially formulated glue that was 'strong enough to maintain the cards in any new position that they might acquire, and yet weak enough to give way slowly to a continued force'. Faraday carefully positioned the bundles around the table, firmly attaching the bottom layer of each one to the tabletop, and drawing a fine pencil line down the edges of cardboard layers. Preparations over, the experiment began. His participants were each asked to place their hands on the top of a bundle, and then have the spirits move the table to the left. After a few moments the table started to shift. Simply by glancing at the bundles he'd prepared, Faraday was able to find the answer to the riddle of table-moving.

It was brilliantly simple. He had reasoned that if a mysterious force was truly acting on the table then the table would move *before* the hands of the sitters did. This would result in the lower layers of each bundle slipping under the upper layers, causing the displaced pencil line to slope from left to right. On the other hand, if the participants' hands were responsible for the table movement then the upper layers of each bundle would move before the lower layers, creating lines that would slope from right to left. When Faraday examined the pencil lines the answer was obvious. Every line sloped from right to left, proving that the participants' hands had moved before the table.

It seemed that Faraday's participants were imagining the table moving and, without realizing it, producing the small hand and finger movements required to make their thoughts a reality. Because these movements were entirely unconscious the twists and turns of the table surprised them, and so were attributed to spirit agency.

Although convinced that he had solved the mystery of table-turning, Faraday realized that Spiritualists might argue that although the unconscious movements of the people around the table were responsible for some of the phenomenon, the spirits were playing a minor but still vitally important role in the movement. The only way of testing this idea would be to eliminate the movement of the hands and see whether the table still turned. Clearly, Faraday couldn't simply ask his participants to stop pushing the table because they had no idea that they were moving it in the first place. A new experiment was required.

Faraday returned to his laboratory and created a second set of ingenious bundles. He now prepared two postcard-sized boards separated by four horizontally placed glass rods that allowed the top board to run freely. This 'board – glass rods – board' sandwich was held together with two large rubber bands. He attached the base of each bundle to the tabletop, and then pushed small metal pins into the sides of the top and bottom boards. Finally, a 15-inch-long stalk of hay was attached vertically to each bundle, with one pin in the lower board and another in the upper.

There was method to his madness. Faraday's design meant that the stalk worked as a lever, with the top pin acting as a fulcrum. Any sideways movement of the top board, no matter how small, resulted in a large and obvious shift in the stalk. The bundles acted as a simple but highly effective way of amplifying the participants' tiny hand movements, and so by asking them to keep the stalk vertical he could ensure that their hands were still.

Faraday brought his merry band of participants together again, and asked them to place their fingers on the top board and attempt to have the spirits move the table, but to try to ensure that the stalk remained vertical at all times. Try as they might, the group simply couldn't budge the table. Faraday correctly concluded that their unconscious movements were completely responsible for the phenomenon, and any consideration of spirit energy was superfluous to requirements.

His findings, published in the *Athenaeum* magazine in 1853, met with a furious response from Spiritualists, with many claiming to be able to produce movement without touching the table at all. They were, however, strangely reluctant to travel to Faraday's laboratory and demonstrate this under controlled conditions.

HOW TO TALK WITH THE DEAD: PART ONE

Running a successful table-turning session is a lesson in applied psychology. To ensure success, try the following ten-step procedure.

1. Choose the right table. Go for something that is about a foot square and two feet high. It doesn't really matter whether it has a round or square tabletop, or is supported by a leg at each corner or on a single pedestal. What does matter is that it tips easily. Test the table by placing your fingertips on the edge of the tabletop and deliberately trying to tilt it over. If it is difficult to budge, find another table.

2. Invite a group of between four and eight people to your house. It doesn't actually matter whether they believe in the afterlife, are agnostic, or completely sceptical. It is more important that they are out to have a good time together.

3. Arrange some chairs in a circle around the table. These seats need to be comfortable, and encourage people to sit forward rather than lean back.

4. Ask everyone in the group to take a seat and place their hands around the tabletop. Their hands do not need to touch their neighbours' hands, and they should rest their fingertips as lightly as possible on the table.

5. Lower the lights, turn on some music and try to establish a light-hearted atmosphere. Ask the group to avoid deliberately pushing the table but instead to focus on keeping their hands as still as possible. Try to get them to chat and joke rather than thinking about obtaining some kind of movement.

6. On a good night you will hear the table starting to creak after around 40 minutes. This is an initial signal that the effect is starting to work.

7. After another ten minutes or so you should get your first movements. If the table is unable to move because it is on thick carpet then it will tip violently and occasionally balance on one or more legs. The group should always try to keep their fingertips in contact with the table, but not prevent any movement. If the table can slide, it may well move around the room. Again, the group should maintain contact with the table and, if necessary, leave their seats and follow it.

8. Do not try to analyse the effect or figure out how it works. Instead, simply enjoy what is happening. Have people remove and replace their hands to see if one person is responsible for the effect. Feel free to ask the table questions and suggest that it answers by tilting or moving in a certain direction. Avoid any possible tears by not suggesting that you have contacted the spirit of someone who was known to a member of the group. Instead, go with contacting a famous, or even fictional, character.

9. If you don't get any creaking or movement after 40 minutes or so, ask everyone to try to will the table to move in a specified direction. It might also be helpful to get the group to try breathing in unison for a minute or so. If you still don't obtain any movement, secretly push the table. This often helps kick-start genuine unconscious movements.

10. At the end of the session, thank the group for participating and tell them that research has shown that the spirits may well follow them home and haunt their dreams for the rest of their lives.

Joseph Jastrow and His Amazing Automatograph

Faraday had shown that small unconscious movements were responsible for table-turning. Inspired by his findings, other researchers explored whether the same type of movements could also account for the curious behaviour associated with the Ouija board.

In my previous book, *Quirkology*, I described the work of one of my academic heroes, a turn of the century American psychologist named Joseph Jastrow. Jastrow carried out many unusual investigations during his career, including work into subliminal perception, the dreams of blind people, hypnosis, and the psychology of magic. However, Jastrow was especially fascinated by the supernatural, and in the 1890s conducted a series of ground-breaking experiments into the Ouija board using a rather strange piece of apparatus called an 'automatograph'.[8]

The principal part of Jastrow's automatograph consisted of two glass plates, each about a foot square, separated by three 'well-turned brass balls'. The bottom plate was attached to the table while the top plate was free to move. Participants placed their hand on the top plate, where even the slightest of hand movements would cause the plate to roll on the balls. To record any movement a pen was attached to the top plate. A sheet of paper, blackened with lamp soot, was placed under the pen, so that any pen movement would be recorded. The paper was then 'made permanent by bathing it in shellac and alcohol'. Like Faraday, Jastrow had constructed a system capable of recording the smallest of movements.

In a long series of experiments, Jastrow hid the recording pen and paper from participants and then asked them to imagine doing three things – making certain movements, looking at different objects around the room, or visualizing a specific part of the room itself. Although the participants didn't realize it, just thinking about a certain direction or location was enough to produce an appropriate movement on Jastrow's glass planchette. Just as Faraday had uncovered the mystery of table-turning, Jastrow had revealed that the same process could account for the movement of the Ouija board. People using such boards were not talking to the dead and communing with the Devil. They were chatting to themselves.

Subsequent research has revealed that these strange movements, known as 'ideomotor' actions, are not confined to table-turning and Ouija boards. In the 1930s, for example, American physician Edmund Jacobson wanted to discover how best to get people to relax.[9] He asked volunteers to think about various subjects while sophisticated sensors monitored the electrical activity in their muscles. When Jacobson asked his participants to imagine lifting their arms the sensors revealed small but real activity in their biceps. Thoughts about lifting heavy weights produced even greater muscle activity. When they were asked to imagine jumping high into the air their leg muscles suddenly showed signs of responding. The phenomenon was not just confined to the body. When the participants imagined the Eiffel Tower their eyes moved up and when asked to recall a poem their tongues moved. Just like Faraday's table-turners 70 years before, Jacobson's participants had no idea that they were making these small movements.

More recent work has shown that these unconscious actions occur regularly.[10] If you think about turning the page of a book, the muscles in your fingers start to move towards the edge of the book. You wonder what time it is and your head begins to look at the clock. You think about making a cup of tea, and your legs kick into action. Although there is some debate as to why these ideomotor actions exist, most researchers believe that they are due to your body preparing itself for the anticipated behaviour. Even a mere thought is enough to make your body put its foot gently on the accelerator and move so that it is better prepared to react when the moment comes.

The scientific study of table-turning and Ouija boards not only provided a solution to these curious phenomena but also resulted in the discovery of a new force of unconscious movement. For more than a hundred years after Faraday and Jastrow's classic experiments researchers believed that talk-ing to the dead was entirely answered by this means. Case closed. Mystery solved. But unbeknownst to them, there was a second, even more intriguing secret hidden within the tipping tables and alphabet cards.

HOW TO TALK WITH THE DEAD: PART TWO

The procedure for a Ouija board session is somewhat similar to table-turning, but has the added advantage of being able to incorporate a test to discover whether the spooky movements are the result of spirit communication or ideomotor action.

1. Choose the right kind of table. This time it needs to be something with a larger tabletop (around two feet square), of normal height but much more sturdy than in the previous experiment. Test the table by trying to deliberately tilt it over. If it is easy to budge, find another table.

2. Write the letters of the alphabet on separate pieces of paper and lay them out in a circle around the edge of the table. Write the word 'Yes' on another piece of paper and 'No' on a final piece. Place these inside the circle of letters.

3. Find a sturdy glass, turn it upside down, and place it in the centre of the circle of letters.

4. Ask everyone to sit around the table to place the first finger of their right hand lightly on the base of the glass.

5. Once again, lower the lights and establish a light-hearted atmosphere. Ask everyone to avoid deliberately pushing the glass and instead to keep their fingers as still as possible. Try to get them to chat and joke.

6. Ask the group to try to contact a spirit. Once again, avoid suggesting anyone known to the group, and instead go with contacting a famous or fictional character. When the glass begins to show signs of movement, ask the spirit to spell their name by moving the glass towards the upturned letters.

7. Once you have established contact and figured out who you are talking to, introduce the notion of a test. Collect the letters of the alphabet, shuffle the pieces of paper, and then deal them *face down* in a circle on the table.

8. Once again have the group ask the spirit to spell out its name. As the glass touches a piece of paper, turn it face up. If the movements of the glass are due to unconscious movement, the selected letters will be meaningless because the group no longer knows where the glass should be heading.

9. If any believers in the group complain that perhaps the message is only meaningless because the spirits can't see the letters either, turn the pieces of paper face up and blindfold the group. Once again, the message should be meaningless.

10. If the group does manage to spell out a name while the letters are face down or they are blindfolded, leave your house immediately and contact your local church for help.

On Trying Not to Think About
White Bears

Many experienced table-turners and Ouija board users rejected the notion of ideomotor action, claiming that the messages from the dead continued to flow thick and fast even when they made a special attempt to keep their fingers completely still. In fact, many reported that they actually obtained even more spectacular results under these conditions. For years scientists attributed these reports to over-active imaginations and the desire to believe, but in the 1990s Harvard psychologist Dan Wegner decided to take a closer look at the claims.

Wegner is a man fascinated by white bears. Or, to be more accurate, he is a man fascinated by asking people not to think about bears. He conducted a series of well-known studies in which he asked participants *not* to imagine a white bear, and to ring a bell each time the unwanted bear sprang into their mind.[11] The results revealed that people had a surprisingly hard time keeping their minds bare of bears, often ringing the bell every few seconds. Wegner had discovered a curious phenomenon known as the 'rebound effect', wherein trying not to think about something causes people to dwell on the forbidden topic. Under normal circumstances people are skilled at distracting themselves and pushing unwanted thoughts out of their minds. However, explicitly ask them not to think

about a topic and they constantly think 'hold on, am I thinking about the thing that I am not supposed to be thinking
about?' and thus are repeatedly reminded about the very
thing that they are trying to forget. Wegner's rebound effect
operates in many different contexts. Ask people to actively
repress unhappy life events and they can't get such thoughts
out of their heads. Ask them to kick stressful thoughts into
touch and they end up becoming especially anxious, and ask
insomniacs to forget about the things that are keeping them
awake and they have an even harder time than usual falling
asleep.[12]

Wegner wondered whether the same phenomenon might
also explain why people were apparently obtaining messages
from tipping tables and Ouija boards despite keeping their
fingers as still as possible. Could the rebound effect also apply
to movement? Could it mean that people who are trying their
very best not to make a certain move are actually more likely
to make the undesired motion?

Wegner decided to carry out an experiment using another
classic example of ideomotor action – the pendulum. For
centuries people have tied small weights to pieces of string
and used the left-to-right or circular movement of the pendulum to try to determine the sex of unborn babies, predict
the future and commune with the spirits. Inviting a group
of participants to his laboratory one at a time, Wegner
positioned a video camera pointing towards the ceiling, and
asked each person to hold a pendulum above it. He asked
half of the participants to make a special effort not to
move the pendulum in a specified direction and the others
to hold the pendulum as still as possible.[13]

The footage from the camera allowed Wegner to carefully

measure the amount of movement in the pendulum. In the same way that being asked not to think about a white bear resulted in endless bears, so trying not to move the pendulum produced increased swinging. These unconscious movements were even more dramatic when Wegner occupied his participants' minds by asking them to remember a six-digit number or count back from 1,000 in threes. These additional findings help explain another curious aspect of table-turning and Ouija boards. Spiritualist lore suggests that the dead are most likely to make their presence known if the people around the table or Ouija board sing hymns, chat or even tell jokes. All of these procedures will tax people's minds and thus be far more likely to encourage people to make unconscious movements.

Wegner's work showed that the rebound effect made table-tipping and Ouija boards especially deceptive. By trying to hold their hands as still as possible and distract themselves from what they were doing, people were creating the perfect conditions for increased ideomotor action and so were especially likely to obtain dramatic effects.

Other work has since shown that this behaviour-based rebound effect occurs in many different situations outside of the séance room. In another study Wegner asked golfers to try to putt a ball onto a spot, and discovered that asking participants not to overshoot the mark made them especially likely to hit the ball too hard. Eye-tracking experiments have revealed that telling football players to avoid kicking a penalty shot into a particular part of the goal resulted in them not being able to keep their eyes off the forbidden area.[14] Athletes have noticed the same effect in real life with, for example, former major league baseball player Rick Ankiel sometimes producing wild throws when attempting to avoid such actions (Ankiel has named the phenomenon 'the Creature').[15] The rebound effect can also affect those trying to change unwanted behaviours, with experiments showing that smokers who try to suppress thoughts about lighting up, and dieters who attempt not to think about fatty foods, find it especially difficult to kick the habit or eat healthily.

Encouraged by his investigations with the pendulum, Wegner turned his attention to the most mysterious of all Spiritualist phenomena – automatic writing. His work was to provide a solution to one of the most taxing philosophical problems of all time.

Mark Twain and the Grand Illusion

Perhaps the most prolific and impressive of all automatic writers was Pearl Curran.[16] Born in 1883 in St Louis, the first 30 years of Curran's life were uneventful, and involved dropping out of high school, trying her hand at various jobs, getting married and teaching music. Then, on 8 July 1913 everything changed. While using a Ouija board to chat with the dead an unusually strong and dominant spirit emerged. The entity explained that her name was Patience Worth and that she had been born in the seventeenth century in Dorset, England, but in later life had taken a ship to America where she was eventually murdered by 'Indians'. Trying her hand at automatic writing, Curran discovered that she could easily channel Ms Worth. In fact, the communications came thick and fast for the next 25 years, with Patience eventually 'dictating' over 5,000 poems, a play and several novels. The quality of the work was as impressive as the quantity. When reviewing Worth's novel about the final days of Jesus, a reviewer at the *New York Globe* favourably compared it to *Ben Hur* while another critic believed it to be 'the greatest story of Christ penned since the Gospels'.

Unfortunately for Spiritualism, Curran's writings failed to provide convincing evidence of life after death. Try as they might, researchers were unable to find any evidence that Patience Worth actually existed, and linguistic analysis of the texts revealed that the language was not consistent with other works from the period. The case for authenticity was not helped by Patience writing a novel set in the Victorian times, some 200 years after her own death. Eventually even the most ardent believer was forced to conclude that Pearl Curran's remarkable outpourings were more likely to have a natural, not supernatural, explanation.

Additional evidence against the spirit hypothesis came from those who claimed to be able to channel famous authors. There's the rather bizarre case of Emily Grant Hutchings, a close friend of Curran, who claimed to be in touch with the spirit of Mark Twain (think 'gravy train'). In 1917 she produced *Jap Herron*, a novel that Hutchings claimed had been dictated to her by the great man himself. Critics were deeply unimpressed, with one noting:

> If this is the best that Mark Twain can do by reaching across the barrier, his admirers will all hope that he will hereafter respect that boundary.

Harper and Brothers, who owned the rights to the work produced by Mark Twain when he was earthbound, took legal action, claiming that the poor quality of *Jap Herron* damaged their sales. As part of their evidence, Harper and Brothers noted that Twain was deeply sceptical about the afterlife and so seemed an especially unlikely candidate as a spirit author. The media had a field day, noting that the Supreme Court would soon have to rule on the issue of immortality. Unfortunately, the case never made it into the courtroom, with Hutchings and her publisher agreeing to withdraw the book from sale prior to the trial.

Assuming that the dead do not have a hand in automatic writing, what are we to make of this curious phenomenon?

Until the mid-1990s by far the most popular explanation involved some form of psychological dissociation. According to this argument, it is possible for some peoples' consciousnesses to become divided into two, with each identity unaware of the other, despite them inhabiting the same brain. It is a strange idea, but nevertheless received widespread support, in part because at the time it was the only show in town. Suddenly everyone and their dog was seen as having multiple personalities and it wasn't long before the idea made it into the world of psychiatry, with clinicians encouraging their patients to experiment with automatic writing as a way of accessing the issues that lay buried deep within their 'subconscious' self.

However, after studying various cases of this strange phenomenon it was again Dan Wegner who advanced a new and radical way of explaining automatic writing. Unlike previous explanations, his idea did not involve the existence of multiple identities trapped within the same skull. Moreover, if he's correct, his work helps solve one of the most hotly debated issues in the history of science.

On the face of it, free will doesn't seem especially contentious. You make a decision to move your wrist and your wrist moves. You decide to lift your leg and up it comes. So far, so what? However, this simple scenario has hidden depths.

Most scientists believe that all of your conscious mental life is the direct result of activity within your brain. For example, right at this moment you are reading the words on this page. Light enters your eyes and triggers cells at the back of your retina. These, in turn, send signals to the visual cortex in your brain, which sets about recognizing the letters and words, and then conveys the required information to the parts of the brain that are able to extract the meaning from the sentences. The process might be extremely complex and difficult to understand, but fundamentally it is all taking place in your eyes and brain.

But when making decisions suddenly, the model doesn't feel quite right. I am going to ask you to make a decision. You can either continue to read this paragraph or go and make a cup of tea. Regardless of your choice, my guess is that it didn't feel like your brain at work. You didn't suddenly feel a rush of blood to the front of your brain, followed by a quick spurt in your left hemisphere. Instead, it felt as if it was 'you', and not a series of electrical impulses in the lump of meat between your ears, that made the decision.

Wegner's neat and clever solution to this mystery involves positing that the sense of 'you' as decision maker is actually a grand illusion created by your brain.[17] According to him, your brain makes every decision in your life including, for example, whether you should stand up, say something or wave your arms around. However, a split second after making each decision your brain does two things. First, it sends a signal to another part of the brain that creates the conscious experience of making the decision, and second, it delays the signal going to your legs, mouth or arms. As a result, 'you' experience the 'I have just made this decision' signal, see yourself act in a way that is consistent with that signal, and incorrectly conclude that 'you' are in the driving seat. In short, you are the ghost in the machine.

THE HELPING HANDS ILLUSION

Many years ago I performed magic on the streets of London's Covent Garden. My act involved selecting a man from the audience and placing a cloak completely around his body. I would then stand behind the man, have him place his hands behind his back, and poke my hands out of two slits in the front of the cloak. To the audience it looked as though the man's hands were poking through the front of the cloak. In reality, they were seeing my hands, not his, and so I could perform tricks and make the man appear to be an expert magician.

Psychologist Daniel Wegner has used exactly the same type of set-up to illustrate another curious aspect of free will. To carry out his demonstration you will need a mirror and a friend. Stand in front of the mirror and have your friend stand behind you. Next, place your hands behind your back and ask your friend to poke their arms under your arms. Now look in the mirror. All being well, your friend's arms will look like your arms (if you are struggling to create this illusion try both wearing black tops). Now have your friend read out the following instructions and then make the appropriate actions with *their* hands

Clench your right hand into a fist three times

Clench your left hand into a fist three times

Wave at the mirror with your right hand

Turn both of your hands palm up and then palm down

Clap your hands together twice

Because your brain finds visual feedback more compelling than movement-related feedback, you should feel as if your friend's hands belong to you and that you are in control of them.

All sorts of clever experiments are put forward by Wegner to support his idea that our feeling of free will is little more than a grand illusion, including one especially curious study conducted by physiologist Benjamin Libet from the University of California in San Francisco in the 1980s.[18]

Imagine travelling back through time and taking part in Libet's experiment. After arriving at his laboratory and having a nice cup of tea, you are taken into a small room and have several small electrodes placed on your head and forearm. Next you are sat in front of a small screen displaying a dot moving in a circle, like the seconds hand of a clock. You are asked to flex your wrist whenever you like, but to report the position of the dot each time you make the decision to flex. After a few wrist flexes the experimenters remove the various electrodes and thank you for your participation.

Like Faraday's study into table-turning, Libet's experiment is as simple as it is ingenious. His experimental set-up measured participants' brain activity, forearm activity and the precise moment that the person thought they decided to move their wrist, allowing him to plot the exact time that each event took place. Libet's data showed a large amount of brain activity about a third of a second *before* each participant said that they made the decision to move their wrist. In short, exactly as predicted by Wegner, your brain appears to make a decision before you are conscious of it.

Libet's experiment is not the only one to suggest that our brain operates before we are aware of it. In the early 1960s neurophysiologist and robotician William Grey Walter asked participants to look at a projection screen and press a button to advance a series of photographic slides one at a time.[19] The participants were connected to various sensors that measured activity in the area of their brain associated with hand movements. Although the participants didn't know it, Grey Walter had hooked the output from these sensors directly to the slide projector to ensure that it was the participant's brain activity, not their button presses, that changed the slides. Exactly as predicted by Wegner's theory of free will, the participants were amazed to discover that the slideshow seemed to predict their decisions.

How does all of this explain automatic writing? Wegner believes that in some people the 'make a decision then create a conscious experience of that decision' mechanism malfunctions. The brain makes the decision to act, and sends the right messages to the appropriate muscles, but fails to send the signals responsible for creating the conscious experience of 'you' making the decision. In automatic writing this results in people scribbling away but with no idea that they are responsible for their jottings. Wegner argues that the phenomenon provides a unique and important insight into the fundamental nature of free will. During such episodes the illusion suddenly breaks down and we are revealed to be the robots that we really are. Automatic writing is not some freak show oddity, but rather reflects the true nature of our everyday behaviour.

Spiritualists were convinced that their techniques for talking to the dead were pushing back the frontiers of science. They were right, albeit for completely the wrong reasons.

These seemingly supernatural phenomena had nothing to do with contacting the spirits, but did yield important insights into the unconscious. Scientific investigations into table-turning and the Ouija board resulted in the discovery of ideomotor action, while similar work with pendulums revealed why people often indulge in the very behaviour that they are trying to avoid. The study of automatic writing played an important role in the development of Wegner's ingenious solution to the age-old philosophical problem of free will. Together, this impressive body of work showed that the unconscious plays a far bigger role in determining behaviour than was previously thought. Merely thinking about any type of activity causes your unconscious mind to automatically and immediately prepare your body to act. By trying not to behave in a certain way you interfere with the usually efficient way that your unconscious controls your actions. And the feeling of free will that you are experiencing right now may well be nothing more than a grand illusion. Ideomotor movements allow you to act in the blink of an eye, the 'rebound effect' has helped explain why many people struggle to quit smoking and lose weight, and Wegner's solution to the free will problem suggests that your brain makes up its mind a fraction of a second before you think that you have made a decision. And all because two young girls once tied an apple to a string, secretly bounced it on the floor, and fooled the world into thinking that it was possible to talk with the dead.

It would be nice to think that modern minds would not be fooled by the ideomotor effects behind table-tipping, Ouija boards and pendulums. Nice, but wrong. Several companies recently claimed to have developed a new form of bomb detector, stating that their product could be employed by the police and military to find concealed explosives, narcotics, and weapons. Operators use the device by inserting a substance-specific 'detection card' into a handheld unit and then walking around until the antenna swings towards the target substance. The Iraqi government spent millions of pounds on hundreds of the devices, deploying them at checkpoints to replace time-consuming physical inspections. As with any dowsing rod, the swinging of the antenna was due to unconscious muscle movements, and tests conducted by the American military revealed that the devices were unable to detect explosives. Unfortunately, by then the damage had been done, with hundreds of civilians being killed by bombs that had passed through the checkpoints undetected. In 1853 Michael Faraday concluded his investigation into the science of table-turning by noting that he was somewhat ashamed of his work, wishing that 'in the present age . . . it ought not to have been required'. Over a hundred and fifty years later it seems that his research is as timely as ever.

INTERMISSION

In which we take a break from our journey,
meet the remarkable Mr Harry Price, travel to the
Isle of Man to investigate a talking mongoose
and end up in the High Court.

So far we have discovered how psychic readings yield important insights into who you think you are, how out-of-body experiences reveal the way in which your brain decides where you are, how displays of alleged psychokinesis show that you are not seeing what is right in front of your eyes, and how attempts to talk with the dead demonstrate the power of your unconscious mind. It is time to catch our breath and take a short intermission before we continue our journey.

When I give public talks about the paranormal I am often asked to describe the oddest piece of research that I have ever come across. It is an easy decision. My chosen piece of work didn't result in any major discoveries about human behaviour or the innermost workings of the brain. It did, however, make the front page of newspapers around the world, led to the most bizarre High Court trial in the history of the British legal system, and provide a fascinating insight into the extremes of human gullibility.

So, sit back, relax and enjoy the strangest investigation in the history of supernatural science. Ladies and gentlemen, I give you . . . Gef, the talking mongoose.

'I Am the Eighth Wonder of the World'

There are many places in the world that have a considerable reputation for paranormal activity. The Isle of Man is not one of them. In fact, according to Wikipedia, the most the island can offer is a malevolent spirit who once blew the roof off a church, a ghostly black dog that wanders aimlessly around a local castle, and a couple of fairies. But, as is so often the case with supernatural science, there is much more to the Isle of Man than meets the eye.[1]

In 1916 James Irving made a strange decision. Finding it increasingly difficult to earn a living as a piano salesman in Liverpool, Irving thought it best if he and his wife Margaret made a fresh start as farmers, and promptly bought a small-holding in one of the most isolated and soulless places on earth. 'Cashen's Gap' was a small farm situated on a wind-swept mountainside on the west coast of the Isle of Man. Five miles from the nearest village, the farm had no electricity or running water, and could only be reached after an hour's climb up a slippery and undeveloped track. Living a life that made Jean de Florette's existence look positively luxurious, James and Margaret Irving found the going tough, often sur-viving solely on rabbits provided by the family sheepdog. After two years at Cashen's Gap, Margaret gave birth to her first and only child, Voirrey (Gaelic for 'bitter').

In the winter of 1928, James added wooden panels to the inside of his farmhouse in an attempt to stave off the intense cold weather, leaving a three inch gap between the panels and walls to help with the insulation. On 12 October 1931 he heard some strange animal-like sounds emanating from behind the panels. Thinking that a small animal had become trapped, James set several traps, laid down some poison, and went to bed. The strange noises continued over the next few days, and in desperation, James attempted to flush out the intruder by making dog-like growls. To his surprise and dismay, the mysterious beast growled right back.

James kept a diary of events and described his next move:

> It occurred to me that if it could make these weird noises, why not others, and I proceeded to give imitations of the calls of other creatures, naming these creatures after every call. In a few days' time one had only to name a particular animal or bird, and it instantly gave the correct call. My daughter then tried it with nursery rhymes, and it had no trouble repeating them. The voice is two octaves above any human voice . . . and its hearing powers are phenomenal. It detects the whisper 15–20 feet away, tells you that you are whispering, and repeats exactly what one has said.

The family started to chat to their new housemate and eventually the mysterious creature spilled the beans. He was Gef, a talking mongoose. Perhaps rather gratuitously, Gef explained that he was quite unlike a normal mongoose. Claiming to have been born in New Delhi in 1852, he also boasted that he was 'extra extra clever' and 'the eighth wonder of the world'.

Gef proved an entertaining companion. He would recite nursery rhymes, tell jokes, and converse in several languages. He was also full of surprises. On one July evening in 1934, for example, James made a note in his diary describing how Gef had sung three verses of the Manx National Anthem, 'in a clear and high-pitched voice; then two verses in Spanish, followed by one verse in Welsh; then a prayer in pure Hebrew (not Yiddish); finishing with a long peroration in Flemish'. The Irvings fed Gef bacon, sausages and bananas. In return Gef caught and killed rabbits, leaving their carcases on a nearby rock for collection.

Although talking to Gef was easy, catching sight of him proved surprisingly difficult. Voirrey was the only person to see him properly, later describing him as 'the size of a small rat with yellowish fur and a large bushy tail'. Margaret also claimed to have stroked Gef through a crack in the wall, but was reluctant to repeat the exercise because he had bitten her finger and drawn blood.

News of Gef eventually spread across the island and soon a stream of visitors came knocking, eager to chat with the Irvings' new-found friend. Within a year, word of the wondrous goings-on at Cashen's Gap crossed the Irish channel to mainland Britain, and journalists from across the land made a pilgrimage to the Irvings' remote farmhouse in the hope of catching a glimpse of Gef. In 1932 a reporter from the *Manchester Daily Sketch* was one of the very few who was fortunate enough to interview Gef:

> The mysterious man-weasel has spoken to me today. I have heard a voice which I should never have imagined could issue from a human throat. The people

who claim it was the voice of the strange weasel seem sane, honest and responsible folk and not likely to indulge in a difficult, long-drawn-out and unprofitable practical joke . . . The weasel even gave me a tip for a winner in the Grand National Horse race!

When news of Gef reached America, one theatrical agent instantly offered the Irvings $50,000 for the film rights. The family refused. Regardless, Gef the talking mongoose was conquering the world.

Harry Price:
Ghost-hunter Extraordinaire

I am very fond of Harry Price. In fact, he is something of a hero to me. Working mainly in the 1930s, Price devoted his life to the scientific study of weird stuff, and under the auspices of his 'National Laboratory of Psychical Research', conducted a series of investigations that both delighted the world's media and infuriated believers and sceptics alike. He exposed famous spirit photographers as frauds (mainly double exposures), tested the alleged 'ectoplasm' material-ized by mediums (largely egg white), re-staged an ancient ceremony to transform a goat into a young man (the goat remained a goat), and filmed the great 'Karachi' as he attempted to perform the legendary Indian Rope Trick (actually Arthur Derby from Plymouth manhandling a stiff rope in Wheathampstead, Hertfordshire). However, in my opinion, his finest moment was the testing of Gef.

In 1932, a friend of the Irvings' wrote to Price, describ-ing the queer happenings at Cashen's Gap, and asking him if he 'would like to interview the little beast'. Price wrote to James Irving, and the two of them struck up a friendly correspondence. Irving repeatedly invited Price to the island, but Price was reluctant to make the long trek and instead sent his friend, a military man called Captain James Mc-Donald.

McDonald arrived at Cashen's Gap on 12 February 1932. On his first day at the farmhouse, Gef remained uncharacteristically quiet, and it wasn't until midnight, when McDonald was leaving for his hotel, that he heard that most traditional of Manx greetings, with the mongoose screaming, 'Who is that bloody man?'

The following day Irving explained that Gef had been quite chatty throughout the night, but had, unfortunately, taken an instant dislike to McDonald. Indeed, the mongoose had requested that McDonald would have to shout 'I do believe in you, Gef!' if their relationship was to continue. McDonald complied, and was greeted by what must have been a stony and somewhat embarrassing silence.

Later that day McDonald overheard Voirrey and Margaret talking to Gef upstairs and shouted, 'Won't you come down? I believe in you!'

'No,' shrieked Gef. 'I don't like you!'

Ever the persistent investigator, McDonald started to creep quietly up the stairs but, in a moment of unfortunate clumsiness, slipped on a loose tread and fell noisily back down. Gef promptly vanished and failed to return during the rest of McDonald's time at the farmhouse. McDonald returned to London and filed a full report for Price.

In March 1935, James Irving sent Price a sample of fur that Gef had allegedly plucked from himself. Price excitedly forwarded it to naturalist F. Martin Duncan for analysis, but was disappointed to receive a report stating:

> I can very definitely state that the specimen hairs never
> grew upon a mongoose, nor are they those of a rat,
> rabbit, hare, squirrel, or other rodent. I am inclined to

think that these hairs have probably been taken from
a longish-haired dog.

Price's suspicion fell on the Irvings' sheepdog, Mona. However, sufficiently intrigued by McDonald's report, he decided
to team up with colleague Richard Lambert and conduct his
own on-site investigation. On 30 July 1935 the two intrepid
investigators arrived on the Isle of Man and made the ardu-
ous climb to Cashen's Gap. Arriving late at night, James and
Margaret introduced them to Voirrey ('now a good-looking
girl of seventeen') and everyone sat around a small table in the
dark-panelled dining room waiting for Gef. James explained
that Gef had not been seen for a few days and was being
especially elusive.

Unperturbed, Price and Lambert addressed all four walls
of the room, explaining that they had travelled a great
distance to be there and were therefore entitled to 'a few
words, a little laugh, a scream, a squeak, or just some simple
scratch'. Nothing. The next morning Price and Lambert again
returned to the farmhouse and were given an extensive tour
of the panelling that apparently allowed Gef to skip unseen
from one room to another. Once again, they pleaded with
the self-proclaimed eighth wonder of the world to make an
appearance. Once again, nothing. Eventually, the intrepid
pair of investigators left, unable to determine whether 'they
had taken part in a farce or a tragedy'. Later James Irving
wrote to Price and described how Gef had reappeared on
the evening of their departure and explained that he had
taken 'a few days' holiday'.

In 1936, Price and Lambert described their investigation
of Gef in a now rare volume, *The Haunting of Cashen's Gap:*

A Modern 'Miracle' Investigated. While not explicitly accusing the Irvings of hoaxing the entire affair, Price and Lambert were less than enthusiastic about the case, concluding that only the most credulous of individuals would be impressed with the evidence for Gef.

Many believed that *The Haunting of Cashen's Gap* would put an end to the entire affair. In fact, it gave Gef the talking mongoose an entirely new lease of life in the most unlikely of places – the British High Court.

The Truth, the Whole Truth, and Nothing but the Truth

Richard Lambert, Price's colleague and co-conspirator in the Gef affair, was an influential figure. As well as being founding editor of *The Listener*, he held a key position on the board of the British Film Institute that, at the time, was under the auspices of the BBC. In early 1936, Lieutenant-Colonel Sir Cecil Bingham Levita, a prominent member of the London County Council, was at lunch with the assistant controller of BBC programmes and suggested that Lambert was unfit to be associated with the BFI because he believed in a talking mongoose. When the remarks reached Lambert he issued a writ for defamation of character.

The case went to the High Court on 4 November 1936 before Justice Swift and a specially convened jury. Each member of the jury was issued with a copy of *The Haunting of Cashen's Gap*. Levita denied slandering Lambert, noting that he hadn't uttered the words, and even if he had, they were fully justified. Lambert countered, claiming that the book accurately represented his views and in no way endorsed the reality of Gef or, for that matter, any other talking mongoose. In keeping with his name, Justice Swift quickly found for Lambert, and awarded him substantial damages of £7,500 (equivalent to roughly £350,000 today). At the end of the trial, Lambert triumphantly autographed the jury's copies of his book.

The trial also had two unintended, but important, consequences. During the case it emerged that the head of Public Relations at the BBC had tried to persuade Lambert to drop the action against Levita for the good 'of his position with the corporation'. Questions were subsequently raised in Parliament, with politicians seeing the affair as yet another example of poor management within the BBC. Prime Minister Stanley Baldwin launched an inquiry, which resulted in the organization moving away from being an 'old boys' network' and introduced formal job interviews and more transparent selection processes. Second, the massive media coverage of the case ensured that mongooses became popular pets throughout Britain.

Eventually, Gef simply vanished. In 1970, writer Walter McGraw tracked down Voirrey and interviewed her about the entire affair. Though eager to keep her current location secret, Voirrey insisted that Gef had indeed existed and had chatted to her on a regular basis. She recounted how the clever mongoose had gone away for progressively longer periods of time, and then one day just never showed up again. Gef was not a positive influence on her life, said Voirrey, wistfully adding, 'Gef has even kept me from getting married. How could I ever tell a man's family about what happened?' Voirrey died in 2005.

In 1937, Cashen's Gap was sold to a Mr Graham and the Irvings returned to mainland Britain. Graham never saw or heard Gef. In 1947, the new owner of Cashen's Gap claimed to have killed a strange animal that was neither ferret nor stoat. His claims remained unverified and the pelt was never analysed. Cashen's Gap was demolished in the 1950s, but the mystery of Gef lives on. Gef has his own Facebook page, and one website dedicated to matters paranormal recently suggested that he may have been 'a supernatural entity from either an alternate dimension or an entity comprised of forces we do not quite understand'.

Perhaps the final word in the whole surreal story should go to Gef. James Irving once described how he reprimanded Gef for taking too long to calculate how many pence there were in seventeen and sixpence. The self-proclaimed eighth wonder of the world responded with a suitably enigmatic reply which, for me, sums up the entire affair beautifully:

'My rectophone wasn't working.'

5. GHOST-HUNTING

In which we spend some quality time with an old hag,
discover why poltergeist researchers once shook a house
to pieces, meet the non-existent phantom of Ratcliffe
Wharf, learn how to see a ghost and explore
the psychology of suggestion.

There is an old joke about a University lecturer who asks his class, 'Has anyone here ever seen a ghost?' Fifteen students put their hands in the air. Next, the lecturer says, 'Well, who here has touched a ghost?' This time only five hands go up. Curious, the lecturer adds, 'OK, has anyone actually kissed a ghost?' A young man sitting in the middle of the lecture theatre slowly raises his hand, looks around nervously and then asks, 'I'm sorry, did you say ghost or goat?'

Thankfully, the results from national surveys have yielded more clear-cut findings. Opinion polls from the past 30 years or so have consistently shown that around 30 per cent of people believe in ghosts and that about 15 per cent claim to have actually experienced one.[1] Additional questioning has revealed that these alleged ghostly encounters do not involve white-sheeted figures drifting through walls, women in black bringing death and destruction, skeletons prancing through cemeteries or headless knights clanking their chains. Despite the frequent appearance of such images in ghost stories and horror films, actual apparitions are far more mundane.

A colleague of mine, James Houran, has carried out a great deal of research into the nature of these ghostly experiences. James is an interesting fellow. During the day this mild-mannered statistician works for a well-known internet dating site creating mathematical models that help promote compatibility. By night Houran transforms into a real life ghost-buster, conducting surveys and studies that aim to solve the mystery of hauntings. A few years ago he analysed almost a thousand ghostly experiences to discover what people report when they believe that they have encountered a spirit.[2]

Houran's work revealed that reports of fully fledged apparitions are very rare. In fact, they only account for 1 per cent or so of sightings and when such figures do turn up they usually appear at the foot of a bed as people are either waking up or drifting off to sleep. Such apparitions have an uncanny knack of looking like a normal person, and their ghost-like nature only becomes apparent when they do something impossible, like suddenly vanish or walk through a wall.

So if people are not seeing full apparitions when they encounter a ghost, just what do they experience? Around a third of Houran's reports involve rather fleeting visual phenomena, such as quick flashes of light, odd wisps of smoke or dark shadows that move furtively around the room. Another third involve strange sounds, such as footsteps from an empty room, ghostly whispering, or inexplicable bumps and knockings. The remaining third are a mixture of miscellaneous sensations, including odd odours of flowers or cigar smoke, sensing a ghostly presence, feeling a cold shiver down the spine, doors opening or closing of their own accord, clocks running especially fast or slow, and dogs being unusually noisy or quiet.

For well over a century scientists have attempted to explain these strange experiences. Some firmly believe that their investigations provide compelling evidence of life beyond the grave. Others are equally convinced that these seemingly supernatural sensations have down-to-earth explanations. Their experiments involve an odd mixture of ground-breaking dream research, camping out in haunted houses, vibrating fencing foils, sitting in the dark waiting for God, shaking entire buildings until they fall to pieces and staging large-scale hoaxes.

Our journey into this mysterious world begins with perhaps the most widely reported of all ghostly experiences.

Henry Fuseli and his Emotionless Horse

In 1781 the Swiss oil painter Henry Fuseli created his most famous work. Entitled *The Nightmare* his painting depicts a woman having a terrible dream and the content of her frightening experience. The woman is sound asleep, lying on her back and with her head hanging down from the edge of her bed. A small evil-looking demon sits on her chest and peers out from the canvas. Towards the back of the painting a horse's head with emotionless eyes is seen emerging from a curtain and staring menacingly at the woman.

The Nightmare proved an instant hit when it was first exhibited at London's Royal Academy, quickly achieved worldwide acclaim, and now features on the cover of almost every academic textbook about the paranormal. Fuseli created another version of the painting several years later, but it is generally agreed that this painting lacks the emotional impact of the original, in part because the demon appears to be wearing a Batman mask and the horse looks like it has just won the lottery.

Fuseli's painting depicts perhaps the most frequently experienced of all ghostly encounters; the arrival of the incubus. According to legend, the incubus is a demon who adopts a male form and forces itself upon sleeping women using its unusually large and cold penis (the Arthurian wizard Merlin was allegedly the result of such an encounter). Sitting on the chest of their victim to prevent movement, the incubus goes about its beastly business while other equally demonic creatures stand by the bedside watching. Never ones to miss an opportunity, it is said that such demons can also take the form of a female succubus and seduce sleeping men (albeit presumably without the aid of an unusually large and cold penis). These creatures have been reported in many different cultures. In Germany the demon is referred to as the 'mare' or 'Alpdruck' ('elf pressure'), in Czechoslovakia they are the 'muera', and the French call them the 'cauchemar'.

Although it is easy to believe that nocturnal demonic experiences could have been the height of supernatural sophistication when Fuseli created his paintings, surely they are not still alive and well in the 21st century? In fact, recent surveys suggest that around 40 per cent of people have experienced exactly the same sensations, including waking up and feeling a crushing weight on their chest, sensing an evil presence, and seeing strange figures in the darkness.[3] These episodes are often interpreted as evidence of demons, ghosts, or even an alien abduction. Regardless of the way in which they are perceived, one point is quite clear – even to the modern mind they are a terrifying and unforgettable experience.

For centuries many of those who have come face to face with night-time demons have been convinced that they have encountered hell on earth. It is only in the last fifty years or so that research has revealed the remarkable truth behind these apparitions.

The Incurably Curious Eugene Aserinsky

The year 1951 did not start well for University of Chicago psychologist Eugene Aserinsky.[4] At work, his post-doctoral research into the eye movements of sleeping babies was proving both slow and unrewarding. At home, Aserinsky was facing severe financial difficulties. His family were forced to live in a small, cold apartment and he could only just afford to rent the typewriter that he needed to write up his work. Years later he described the sense of desperation that he faced:

> If I had a suicidal nature, this would have been the time. I was married and had a child. I'd been in universities for twelve years with little to show for it. I was absolutely finished.

In addition, he was exploring avenues that simply didn't interest his more mainstream colleagues. The vast majority of academics at the time assumed that the brain switched off when people drifted into the land of nod and turned back on when they woke up, and didn't share Aserinsky's interest in the psychology of sleep. However, Aserinsky wanted to discover if this 'move on, nothing to see here' approach to the sleeping brain was correct. Unable to attract proper funding for his work, he found an old brainwave measuring machine (referred to as an 'electroencephalograph') in the basement of his department, dragged it up into his office and managed to get it working. Unfortunately, one major problem remained – without proper funding who would be willing to spend several unpaid nights in Aserinsky's sleep laboratory covered in sensors? Eventually he managed to come up with a lateral solution to this problem as well. On a cold evening in December 1951 he tucked his eight-year-old son Armond into the laboratory bed, connected eye movement and brainwave sensors to Armond's face and head, and retreated to his office.

After an hour or so, Armond drifted off to sleep and the experiment began. For the first 40 minutes or so Aserinsky carefully monitored the pens tracing the output from the electroencephalograph. The lack of movement was under-whelming, and it looked like the scientific establishment had been right to let sleeping brains lie. About 20 minutes later the pens started scribbling away, indicating large amount of activity from both the eye movement and brain activity sen-sors. Assuming that his son had woken up, Aserinsky went to check that he was okay. When he opened the door to his laboratory, he couldn't believe his eyes. His son was sound asleep.

At first Aserinsky assumed that his experimental equip-ment was faulty and set about checking the large number of leads going in and out of the electroencephalograph. No obvious problems emerged. The following day he showed the charts to his supervisor who also thought that there must have been a problem with the equipment, and asked Aserinsky to run a second set of more thorough checks. The system came back with a clean bill of health. A few more nights of monitoring Armond in the sleep laboratory convinced Aserinsky that his findings were genuine. At certain points during the night the sleeping brain became mysteriously and amazingly active. Additional work revealed that these sudden rushes of brain activity were accompanied by rapid eye movements or, as Aserinsky referred to them, 'REMs' (he originally wanted to call these 'jerky eye movements' but was worried about the negative connotations of the word 'jerk'). Not only that, but whenever Aserinsky woke a participant up after a period of REM the person almost always reported a dream.

In September 1953 Aserinsky and his supervisor published their findings in a now-classic paper entitled 'Regularly Occurring Periods of Eye Motility, and Concomitant Phenomena, During Sleep', and changed psychology for ever.[5] Suddenly researchers realized that there was a great deal more to the sleeping brain than they had previously assumed, and that Aserinsky had discovered a way of entering the hitherto hidden world of dreaming. As one researcher later remarked, it was like discovering 'a new continent in the brain', and scientists across the world were suddenly eager to explore this brave new world. Strangely enough, Eugene Aserinsky didn't join them. Ever the unconventional but incurably curious polymath, he left the University of Chicago soon after his ground-breaking experiment to investigate the effects of electrical currents on salmon.

To Sleep, Perchance to Dream –
Ay, There's the Rub

Researchers have now identified five distinct stages of sleep.

Soon after nodding off you drift into the creatively labelled 'Stage 1'. Here your brain is still very active and producing high frequency brain waves known as 'Alpha' waves. During this stage people frequently experience two types of halluci-nations known as hypnagogic imagery (which occur when people are drifting into sleep) and hypnopompic imagery (which occur when they are waking up). Either type can result in a wide range of visual phenomena, including random speckles, bright lines, geometric patterns, and mysterious animal and human forms. These images are often accom-panied by strange sounds such as loud crashes, footsteps, faint whispers, and snatches of speech. Interestingly, these are exactly the type of experiences that have been mistaken for the presence of a ghost for hundreds of years.

Having survived the potential terrors associated with 'Stage 1' you drift into 'Stage 2'. Again, your brain is far from calm, often producing brief bursts of activity known as 'spindles'. 'Stage 2' lasts for about 20 minutes and can result in the occasional mumble and even full on sleep-talking. Slowly you drift further down into, you guessed it, 'Stage 3'. Now your brain and body are starting to become properly relaxed and after another 20 minutes or so you finally enter the deepest stage of sleep . . . In 'Stage 4', your brain activity is at a minimum, resulting in very slow moving 'Delta' waves. If you are going to engage in a spot of bedwetting or sleep-walking, this is the moment.

After around 30 minutes or so in 'Stage 4' something very strange happens. Your brain moves rapidly back through the first three stages and then enters a mysterious state. It exhibits the same high levels of activity originally displayed during 'Stage 1', but your heart races, your breathing becomes shal-low, and you produce the REMs that so fascinated Aserinsky all those years ago. Now you are dreaming. Everyone experi-ences this REM stage about five times each night, with each of the periods lasting an average of twenty minutes. Although some people think that they don't dream, if they are woken up directly after exhibiting REMs, more often than not, they will report a dream. It is not that some individuals don't dream, but rather that they don't remember their dreams in the morning.

Additional work has shown that two curious things hap-pen to your body when you dream. First, your genitals become active, with men getting an erection and women exhibiting increased vaginal lubrication. Although hailed as a breakthrough in the 1960s, some researchers have argued that the effect may have been discovered long before, pointing out, for example, that one of the 17,000-year-old cave paint-ings in Lascaux depicts a dreaming Cro-Magnon hunter with an erect penis (then again, he might just have really enjoyed hunting). Second, although your brain and genitalia are very active during dreaming, the rest of your body is not. In fact, your brainstem completely blocks any movement of your limbs and torso in order to prevent you acting out your dreams and possibly hurting yourself.

Just as your brain can fool you into seeing an afterimage of a ghost, it can also trick you into thinking that you have encountered an evil entity. As you move between 'Stage 1' and the REM state your brain sometimes becomes confused, causing you to experience the hypnagogic and hypnopompic imagery associated with 'Stage 1', but the sexual arousal and paralysis associated with the REM state. This terrifying combination causes you to feel as if a heavy weight is sitting on your chest and pinning you to the bed, sense (and sometimes see) an evil entity or two, and believe that you are having a rather odd form of intercourse.

For centuries a significant percentage of the public have convinced themselves that they have been attacked by demons, ghosts and aliens. Not only have sleep researchers revealed the true nature of such experiences, but also uncovered the best way of banishing these entities from your bedroom. Perhaps not surprisingly, this does not involve extensive chanting, the sprinkling of holy water or an elaborate exorcism. In fact, it turns out that all you have to do is try your best to wiggle a finger or blink. Even the smallest of movements will help your brain shift from the REM state to 'Stage 1' of sleep, and before you know it you will be wide awake and safely back in the land of the living.

Those who believe in ghosts have now been forced to accept that the incubus experience is not evidence of hell, but rather a clever trick of the mind. However, rather than jettison their belief in hauntings, they have focused their attention on an altogether trickier problem – the many ghost sightings that happen when people are far from sleep.

HOW TO SUMMON THE SPIRITS

Would you like to see a ghost right now? If so, stare at the small white dot in the left-hand box below for about thirty seconds, and then look at the small black dot in the empty right-hand box. After a few moments you should see a mysterious woman in white emerge in front of your eyes. If you repeat the exercise, but look at a white wall rather than the tiny box, you will see a giant ghost projected onto the wall.

Psychologists refer to the ghostly figure that you just saw (and that many of you will continue to see for the next few minutes – sorry about that) as an 'afterimage'. Your perception of colour is based on three systems. Each of these systems is based around two colours, with one dealing with the red-green continuum, another with blue-yellow and the third with black-white. In each of these systems the two colours oppose one another and can't be seen together. For example, when the eye and brain encounter the colour red, the 'red' half of the red-green system is activated, disabling your ability to see anything green at the same time (this explains why you never see colours that appear yellowish-blue or reddish-green).

When you stared at the solid black image a few moments ago you unwittingly forced the 'black-white' neurons to quieten down for a very long period of time. Then, when you shifted your attention to the empty box, the neurons became activated. However, because they were already in a quiet state, the activation made them become over-excited, creating a rebound effect that resulted in a white afterimage.

The Rose Without a Thorn

Hampton Court Palace has a long and controversial history. In the early 1500s the Archbishop of York, Cardinal Thomas Wolsey, invested seven years of his life and over 200,000 gold crowns building a palace fit for a king. A few years after completing the project Wolsey fell out of favour with the reigning monarch, Henry VIII, and felt it would be politically expedient to gift his beloved palace to the Royal family. Henry graciously accepted Wolsey's kind offer, expanded the estate to ensure that it could hold his thousand-strong court, and promptly moved in. The palace went on to become home to some of Britain's most famous kings and queens before being opened to the public in the mid-nineteenth century. Nowadays Hampton Court Palace is one of Britain's most popular historical attractions, playing host to more than half a million visitors each year.

The palace is famous for many things. It houses invaluable works of art from the Royal Collection, contains the best preserved medieval hall in Britain, and boasts giant Tudor kitchens designed to feed 600 twice a day. Oh, and one other thing. It is also one of the most haunted buildings in Britain. Various spirits allegedly haunt the palace. There is, for example, a 'lady in grey' who walks through the cobbled courtyards regular as clockwork, a 'woman in blue' who continuously searches for her lost child, and a phantom dog that lives in Wolsey's closet. However, despite stiff competition,

Hampton Court's most famous spirit is that of Catherine Howard.

Henry VIII did not have a great track record when it came to relationships, of course. He cheated on his first wife, beheaded his second, lost his third while she was giving birth to his only son, and divorced his fourth. In a move that would make even the most experienced marriage counsellor raise an eyebrow, the 49-year-old Henry then became infatuated with a 19-year-old courtier named Catherine Howard. After a brief period of wooing Henry married Howard, publicly declaring that she was his 'rose without a thorn'.

A few months after getting married, Howard found herself very much in love. Unfortunately, the apple of her eye was not her husband Henry, but rather a young courtier named Thomas Culpepper (who, according to several accounts more than lived up to his reputation as a 'gentleman of the bedchamber'). News of their affair eventually reached Henry, who promptly decided to fetch the garden shears and remove the head of his beloved rose. Upon hearing the bad news, Catherine was understandably upset, and ran to Henry to plead for her life, but was stopped by Royal guards and dragged back through the corridors of the palace to her apartments. A few months later both Thomas Culpepper and Catherine Howard were beheaded at the Tower of London.

The ghost of Catherine Howard is said to haunt the corridor that she was dragged down against her will. By the turn of the last century this area of the palace had become associated with a whole host of ghostly experiences, including sightings of a 'woman in white' and reports of inexplicable screams.

In January 2001 a palace official telephoned me, explained

that there had been a recent surge in Howard-related phenomena, and wondered whether I might be interested in investigating.[6] Eager to use the opportunity to discover more about hauntings I quickly put together an experiment, assembled a research team, photocopied hundreds of blank questionnaires, loaded up my car and headed off to the palace for a five-day investigation.

The palace had called a press conference to announce the start of my study, and had attracted the attention of journalists from all around the world. We decided to make the press conference a two-part affair, with a palace official talking about the history of the haunting in the first half, a brief break, and then my good self describing the forthcoming investigation. A palace historian kicked off the proceedings by telling a packed room of reporters what happened when Henry met Cathy. During the brief break I stepped outside to get some fresh air and the strangest thing happened. A car containing two tipsy teenagers drove slowly past me. One of the teenagers wound down the window and threw an egg at me. The egg smashed on my shirt. Unable to change, I tried to remove the worst of the stains and then returned to the press conference. A few minutes into my talk one of the journalists noticed the marks on my shirt and, assuming that it was ectoplasm, asked whether Catherine Howard had already slimed me. I replied 'Yes. This is going to be a tougher investigation than I first thought.' Although said in jest, my comment was to prove prophetic.

Prior to the experiment, I had asked the palace to supply me with a floor plan of the corridor that would have held such unpleasant memories for Catherine Howard. I then met with Ian Franklin, a palace warder who had carefully catalogued a century of reports of unusual phenomena experienced by staff and visitors, and asked him to secretly place crosses on the floor plan to indicate where people had consistently reported their experiences. To avoid any possible bias during the investigation, neither I, nor any other member of the research team knew which areas had been marked by Ian.

During the day groups of visitors were transformed into ghost hunters. After hearing a brief talk about the project, each participant was handed a blank floor plan, asked to wander along the corridor and place an 'X' on the floor plan to indicate the location of any unusual experiences that they might have (essentially playing a game of 'spot the ghoul'). Each night we would place a variety of sensors and a £60,000 heat imager in the corridor in the hope of catching Catherine mid-'boo'.

Day one of the investigation went badly, with several participants wandering into the wrong corridor and then wondering why the floor plan was so wildly inaccurate. On day two we were joined by a woman who claimed to be the reincarnation of Catherine Howard, and said that she could provide a unique first person perspective on the proceedings ('Actually, I was dragged up the corridor, not down it', 'Not sure that the new paint job in the kitchens works for me', etc). On day three a Brazilian film crew attempted to film in the haunted corridor but the presenter suddenly had an anxiety attack and left the palace without completing the piece. Day four turned out to be especially interesting. The team (which now included the reincarnated Catherine Howard) assembled in the morning as usual and reviewed the heat sensor data from the previous night. It was immediately obvious that something very strange had taken place, with the graphs showing a massive spike in temperature around 6 a.m. We eagerly rewound the recording from the thermal imager to discover if we had caught Catherine on tape. Dead on 6 a.m. the doors at one end of the corridor burst open and in walked a figure. The reincarnated Catherine Howard instantly recognized the figure as a member of Henry VIII's court. However, a few seconds later the proceedings took a decidedly more sceptical turn when we saw the figure walk over to a cupboard, remove a vacuum cleaner and start to clean the carpets. Thankfully, the data from the rest of the investigation proved more revealing.

Field footage of thermal 'ghost'
www.richardwiseman.com/paranormality/ThermalGhost.html

First of all, people who believed in ghosts experienced significantly more strange sensations than the sceptics. Interestingly, these odd experiences were not randomly spread throughout the corridor but rather stacked up in certain areas. Even more interestingly, these areas corresponded to the ones that Ian Franklin had identified from analysis of previous reports. Given that neither the team nor volunteers knew the location of these areas during the study, it was strong evidence that something strange was happening.

We have obtained the same pattern of findings in several

investigations at other haunted locations. Time and again those who believe in the paranormal experience more ghosts than those who don't, and these sensations frequently occur in places that have a reputation for being haunted. As I loaded my equipment back into my car and said goodbye to our well-meaning but intensely annoying Catherine Howard wannabe, one question nagged away in my mind. Why?

The Machine in the Ghost

Spend any time looking at websites about ghost-hunting or reading books about hauntings and you will soon come across the 'Stone Tape Theory'. According to its proponents, ghosts are the result of buildings recording and then replaying past events. To put it another way, ghosts don't just walk through walls but are actually part of them. The idea has emotional appeal but, from a scientific perspective, suffers from three significant problems. First, the idea is quite literally a work of fiction. In December 1972 the BBC broadcast a Christmas ghost story entitled *The Stone Tape*. Written by Nigel Kneale (who also penned the fabulous *Quatermass*), the play centres on a group of scientists investigating an old haunted house. The researchers discover that the stone in one of the rooms is capable of recording past events, and that the alleged ghosts are actually these recordings being replayed. Curious to discover more, the team carry out various experiments and (as is often the way when fictional scientists meddle with the unknown) unwittingly release a malevolent force into the world. The second problem with the theory is that it is completely implausible – as far as we know, there is no way that information about events can be stored in the fabric of a building. And the third and final problem – and from a scientific perspective this is perhaps the biggest stumbling block – is that there is not a shred of evidence to suggest that it is true.

Thankfully, other scientists have come up with more plausible ways of explaining things that go bump in the night. In the 1950s Mr G. W. Lambert, the president of the Society for Psychical Research, suggested that the answer lay not in the walls of haunted buildings but rather in the natural movement of earth and water deep beneath their foundations.[7] Lambert speculated that the rise in underground streams following heavy rainfall could induce structural movement in a house that could, in turn, cause doors to creak and objects to move around. Eager to test Lambert's theory, Cambridge researchers Tony Cornell and Alan Gauld carried out one of the most bizarre, and often overlooked, studies in the history of ghost-hunting.[8]

BOO!

Psychical researcher Tony Cornell carried out a great deal of fascinating work into the unknown, but perhaps his strangest series of studies aimed to assess the reliability of eyewitness testimony for ghosts.[9] The idea was simple. First, Cornell and his colleagues would dress up as apparitions, stand in various public spaces at night, and attract the attention of passers-by. Next, other members of the research team would interview these eyewitnesses and assess the accuracy of their testimony. However, as is often the case with supernatural science, the studies proved surprisingly difficult to conduct.

In their initial experiment Cornell wrapped himself up in a white sheet and spent several nights walking around a dark park in the centre of Cambridge. Although 80 people were in a position to see the fake spirit, none of them appeared to notice the strange goings-on. Wondering if the disappointing results were due to poor illumination Cornell put the sheet on again and spent several nights walking around a well-lit Cambridge graveyard. A total of 90 cars, 40 cyclists and 12 pedestrians passed by, but only four people appeared to notice the apparition. Of these, two were interviewed, with one saying that he had assumed that the 'ghost' was part of an art project and the other remarking that the person under the sheet 'surely must be mad'. In a final attempt to be spotted, Cornell contacted a local cinema and arranged to re-stage his ghost walk in front of the screen just prior to the showing of an X-rated film (chosen 'to safeguard against children being present'). The audience were then asked to raise their hands if they had seen something unusual, revealing that a third of the audience had failed to notice the fake spirit. The testimony from those who did spot the figure was often far from accurate, and included a description of a young girl dressed in a summer frock, a woman dressed in a heavy coat, and a polar bear ambling across the screen.

Cornell's findings suggest that if the dead do indeed walk among us they might benefit from wearing a high-visibility vest.

Gauld and Cornell found a house that was scheduled for demolition and persuaded the local council to give it to them for the purpose of serious scientific research. The duo started off by cementing a powerful vibrating machine to the wall of the house. Next they slung a long rope around the chimney and attached a heavy weight to the end of the rope. They then ventured inside the house and carefully positioned 13 'test' objects in different rooms, for example, placing a marble on the floor in one room and a teacup and saucer on a shelf in another. Preparations complete, they moved onto the second stage of the experiment.

Gauld positioned himself inside the house and Cornell switched on the giant vibrator. The entire house shook but none of the test objects moved an inch. Cornell then arranged for the heavy weight on the end of the rope to be lifted and slammed into the side of the building. All the test objects remained unmoved by the experience. The following day Gauld and Cornell returned to the house, turned the vibrator up to 11, and finally managed to get the teacup to rotate in the saucer. The dynamic duo then repositioned the vibrator for even greater effect and took up positions in the house for one final test. As a colleague turned the vibrator's dial to maximum Gauld and Cornell felt the entire house shake. Dirt came crashing down the chimney, slabs of plaster fell from the ceiling, and a large crack emerged in one of the bedroom walls. Subsequently describing their time there as 'quite our most terrifying experience in pursuit of a poltergeist', they stood their ground and observed that even under these extreme conditions only a few of the test objects moved (a plastic beaker fell over, the cup and saucer fell off the shelf and a plaster of Paris donkey moved a fraction of an inch away from the wall). After putting their lives on the line in the pursuit of scientific knowledge, Gauld and Cornell concluded that Lambert's theory simply didn't hold water.

Lambert is not the only one to suggest that hauntings might be the result of bad vibrations. In my previous book, *Quirkology*, I described another idea put forward by electrical engineer Vic Tandy.[10] In 1998, Tandy was working in a laboratory that had a reputation for being haunted. Working alone in the lab late one August night, he started to feel increasingly uncomfortable and had the distinct impression of being watched. As he slowly turned around he saw an indistinct grey figure slowly emerge from the left side of his peripheral vision. With the hair on the back of his neck standing to attention, Tandy eventually built up the courage to look directly at the figure. As he did, it faded away and disappeared.

Tandy was a keen fencer and the following day brought his foil into the laboratory for repairs. As he clamped the foil into a vice, it started to vibrate frantically. Although initially baffled, he eventually figured out that the air conditioning unit in the room was producing a low frequency sound wave that fell well below the human hearing threshold. These waves, referred to as 'infrasound', vibrate at a frequency of around 17Hz, and are capable of producing weird effects. Tandy speculated that in some allegedly haunted buildings certain naturally occurring phenomena, such as strong winds blowing across an open window or the rumble of nearby traffic, could be creating infrasound and giving people strange experiences that they incorrectly attribute to the presence of spirits.

There is some evidence to support Tandy's idea. For example, in 2000 he reported investigating a fourteenth-century cellar in Coventry that had a reputation for being haunted, and found infrasound in the part of the cellar where many people had reported seeing apparitions.[11] As I also noted in *Quirkology*, some additional research has suggested that people do have strange experiences when exposed to low frequency sounds. However, although the theory might explain some alleged ghostly activity, the required combination of strong winds, specifically shaped windows and nearby traffic mean that it is unlikely to account for a large number of hauntings.

Of course, as a scientific explanations for spirits, infrasound is not the only show in town. . . .

Waiting for God

Neuropsychologist Michael Persinger, from Laurentian University in Canada, believes that ghostly experiences are caused by the brain malfunctioning and, more controversially, that these sensations can be easily elicited by applying very weak magnetic fields to the outside of the skull.[12]

In a typical Persinger study participants are led into a laboratory and asked to sit in a comfortable chair. They then have a helmet placed on their heads, are blindfolded, and are asked to relax for about 40 minutes. During this time several solenoids hidden in the helmet generate extremely weak magnetic fields around the participant. Sometimes these fields are focused over the right side of the head, at other times they switch to the left and once in a while they circle around the skull. Finally the helmet and blindfold are removed and the participant is asked to complete a questionnaire indicating if they experienced any strange sensations, such as the sense of a presence, vivid images, odd smells, being sexually aroused or coming face to face with God.

After years of experimentation, Persinger claims that around 80 per cent of participants tick the 'yes' box to at least one of these experiences, with some even going for the 'all of the above' option. The study has featured in lots of science documentaries, resulting in several presenters and journalists putting Persinger's magic helmet on their heads in the hope of meeting their maker. For the most part, they have not been disappointed. Parapsychologist Sue Blackmore felt as if something had got hold of her leg and dragged it up the wall, followed by a sudden sense of intense anger (which is exactly how I would feel if someone took my leg and dragged it up a wall). *Scientific American* columnist and sceptic of the paranormal Michael Shermer had an equally strange time under the influence of the helmet, feeling a strange presence rush past him, followed by a sense that he was drifting out of his body. Persinger does not, however, have a 100 per cent track record, with evolutionary biologist and well-known atheist Richard Dawkins feeling very little, followed by a strong sense of disappointment.

Despite the occasional unresponsive atheist, all was going well with Persinger's theory until a team of Swedish psychologists, lead by Pehr Granqvist from Uppsala University, decided to carry out the same type of experiments.[13] It all started well, with some of the Swedes visiting Persinger's laboratory and even borrowing a portable version of one of his helmets for their own study. However, Granqvist became worried that some of Persinger's participants may have known what was expected of them and their experiences could therefore have been due to suggestion rather than the subtle magnetic fields. To rule out this possibility in his own work, Granqvist had all of his participants wear Persinger's borrowed helmet, but ensured that the coils were only turned on for half of the participants. Neither the participants nor the experimenters knew when the magnetic fields were on and when they were off.

The results were remarkable. Granqvist discovered that the magnetic fields had absolutely no effect. Three of his participants reported intense spiritual experiences, but two of these were not being exposed to the magnetic fields at the time. Likewise, 22 people reported more subtle experiences, but 11 of them were in the 'coils off' condition. When Granqvist's work was published in 2004, Persinger argued that the poor showing may have been due, in part, to the participants in the 'coils on' condition only being exposed to the magnetic fields for 15 minutes, or Granqvist running the DOS-based software controlling the coils in Windows and thus possibly altering the nature of the magnetic fields. The Swedish team defended their work and stood by their findings.

Worse was to come for Persinger. In 2009, psychologist Chris French and his colleagues from Goldsmiths College in London carried out their own investigation into Persinger's ideas by hiding coils behind the walls of a featureless white room, and then asking people to wander around the room and report any strange sensations.[14] Seventy-nine people visited this most scientific of haunted houses for about 50 minutes each. Following in the footsteps of Granqvist, French and his team ensured that the coils were only switched on for half of the visits, and that neither the participants nor experimenters knew whether the coils were on or off. The magnetic fields had absolutely no effect on whether or not people reported a strange experience.

Some commentators have noted that we are all subjected to far greater magnetic fields whenever we use a hairdryer or turn on a television set, and so, if the theory worked, we would experience ghosts far more frequently.

The idea of infrasonic ghosts and electromagnetic spirits has caught the imagination of the media and public alike. However, the scientific jury is unconvinced.

GHOST-HUNTING

So has anyone solved the mystery of hauntings? Before we delve deeper, it is time to discover more about the spectre of a rather strange clerical ghost.

The Power of Raman Spectroscopy

A few years ago I conducted an unusual experiment as part of a television series on human behaviour. We assembled 20 unsuspecting volunteers in a room, had them sit in four rows of chairs and explained that we were about to test their sense of smell. They were shown a small perfume bottle containing bright green liquid and we explained that once the lid of the bottle was unscrewed a strong peppermint smell would permeate throughout the room. We then carefully removed the lid and asked people to raise their hands once they could smell the peppermint. Within moments a few people in the front row raised their hands. Seconds later those in the second row followed suit. Before long about half of the group had their hands in the air. When we asked people to describe the smell they said that it was fresh, pleasant and stimulating. There was just one small problem. As you might have guessed by now, the bottle actually contained a mixture of water and odourless dye. The peppermint smell existed solely in the minds of the participants and was designed to demonstrate the power of suggestion.

This demonstration, first conducted by Edwin Emery Slosson in 1899 (who, according to a report from the time, was 'obliged to stop the experiment, for some of the front row seats were being unpleasantly affected and were about to leave the room'), has been carried out in psychology departments across the land for over a hundred years.[15]

In the late 1970s sensory scientist Michael O'Mahony from the University of California took the idea to new heights when he persuaded the BBC to undertake an ingenious version of the study during a live programme.[16] O'Mahony constructed some mock scientific apparatus (think weird-looking large cone, masses of wires and several oscilloscopes), and managed to keep a straight face as he told viewers that this newly devised 'taste trap' used 'Raman Spectroscopy' to transmit smells via sound. He then proudly announced that the stimulus would be a country smell. Unfortunately, the studio audience interpreted his comments to mean 'manure', resulting in a significant amount of smutty laughter. After clarifying that they would not be broadcasting the smell of shit into people's homes, the research team played a standard Dolby tuning tone for ten seconds. Just as the bottles in the more pedestrian versions of the study contained nothing but water, so the tone did not actually have the ability to induce smells.

Viewers were then asked to contact the television station and describe their experiences. A few hundred viewers responded, with the majority stating that they had detected a strong smell of 'hay', 'grass' and 'flowers'. Although they were explicitly told that the smell would not be manure-related, several people mentioned that they had detected the subtle hint of silage. Many respondents described how the tone had brought about more dramatic symptoms, including hay fever attacks, sudden bouts of sneezing and dizziness.

These experiments demonstrate how nothing more than the power of expectation can cause some people to experience various smells. James Houran (of internet dating and ghost-busting fame) also believes that they play a vital role in unlocking the mystery of hauntings.

Houran speculated that if suggestible people believe that they are in a haunted house, they may experience the strange sensations typically attributed to ghostly activity. In addition, he noted that those experiences are likely to create a feeling of fear that will cause people to become hyper-vigilant and pay attention to the subtlest of signals.[17] They will suddenly notice that tiny creak in the floorboards, the swaying of the curtains, or a brief whiff of burning. All of this will cause them to become even more afraid and therefore exhibit even greater hyper-vigilance. The process feeds on itself until the person starts to become highly agitated, anxious and prone to more extreme sensations and hallucinations.

The findings from many studies support Houran's ideas. In my own work, those who believed in ghosts reported far more weird experiences than sceptics, and their sensations tended to focus around the type of scary-looking locations that frequently feature in horror films. In the experiments investigating the (lack of) impact of weak magnetic fields on the brain, those reporting strange experiences tended to be far more suggestible than most. Although these findings are encouraging, the ultimate testing of the theory involves taking suggestible people to a place that does not have a reputation for being haunted, making them believe that it does, and seeing if they experience the same kind of ghostly activity reported in 'genuine' hauntings. Houran has conducted several of these experiments, with intriguing results.

In one experiment he took over a disused theatre that had absolutely no reputation for being haunted, and asked two groups of people to walk around it and report how they felt.[18] Houran told one group that the theatre was associated with lots of ghostly activity and the other that the building was simply undergoing renovation. Those in the 'this building is haunted' group reported weird sensations all over the place, while the other group experienced nothing unusual. In another study Houran asked a married couple living in a house that had no reputation for ghostly activity to spend a month making a note of any 'unusual occurrences' that they noticed in their home.[19] Reporting the results in the paper 'Diary of events in a thoroughly unhaunted house', he noted that the couple reported an amazing 22 weird events, including the inexplicable malfunctioning of their telephone, their name being muttered by a ghostly presence, and the strange movement of a souvenir voodoo mask along a shelf.

Although these studies are impressive, perhaps the prize for the best test of Houran's theory goes to journalist Frank Smyth.

The Phantom Vicar of Ratcliffe Wharf

In 1970 Frank Smyth was an associate editor of a magazine dealing with paranormal phenomena known as *Man, Myth and Magic*.[20] One Sunday morning Smyth travelled to Ratcliffe Wharf in London's Docklands to meet his friend John Philby (son of spy Kim Philby). Throughout the nineteenth century Ratcliffe Wharf was a busy dock. As a result of the constant coming and going of sailors, it also became a hotbed of iniquity, crammed full of gambling dens, drinking houses and brothels. Philby was renovating an old warehouse in the area, and suggested to Smyth that it might be fun to create a ghost story.

After a couple of hours of productive brainstorming in a nearby pub, Smyth and Philby emerged with the phantom vicar of Ratcliffe Wharf – an emotive tale of sex, sailors, and murder. So, if you are sitting comfortably, I will begin . . .

In the early 1800s a former vicar of the Wharf's largest church, St Anne's, set up a guesthouse in the area for sailors. However, when business failed to boom this most corrupt of clergymen adopted more unsavoury ways of making ends meet. The vicar would pay young attractive women to tempt sailors back into his guest house, ply the men with drink, and then invite them upstairs for a bit of 'how's your father?' When the men stripped off and climbed into bed, the vicar would emerge from his hiding place in the room, batter them to death with his silver-topped cane, steal their money and dump their lifeless bodies into the muddy Thames. According to local lore, the vicar's spectre still haunts the area.

After carefully checking that the area was not associated with any ghostly activity, Frank described his completely fictitious story in the latest issue of *Man, Myth and Magic*, noting that both he and Philby had actually seen the ghost.

Three years later a BBC documentary programme described the hoax, presenting a dramatized account of the phantom vicar of Ratcliffe Wharf (including a sign outside the guesthouse where the vicar used pretty women to lure sailors, appropriately announcing 'Lodgings for Seamen') and went in search of people who had seen the non-existent ghost. They didn't have to look far. One local woman reported seeing the ghostly vicar and described how he was dressed in a white shirt, with a cloak and flowing grey hair. Believing the ecclesiastical spirit to be a rather lecherous figure, the woman described how she often had the sense that he was watching her when she undressed at night. Next, a landlord in the area described how his daughter and her two-year-old son had had a chilling encounter with the spectre when they had come to stay with him. After several sleepless nights the child pointed to an area of the room and screamed out that he didn't like the man standing there. The child's mother then turned around and saw the ghostly vicar looking at her. Other witnesses included a workman who had seen the vicar walk round a corner before melting away in front of his eyes, and two policemen who told the untruth, the whole untruth and nothing but the untruth about the ghostly activity in the wharf.

The phantom vicar of Ratcliffe Wharf is a vivid vindication of Houran's theory. Hauntings do not require genuine ghosts, recording walls, underground streams, low frequency sound waves or weak magnetic fields. Instead, all it takes is the power of suggestion.

The Big Question

Although the psychology of suggestion accounts for many ghostly phenomena, there still exists one final mystery – why on earth should our sophisticated brains have evolved to detect non-existent ghostly entities?

Scientists have proposed various theories to account for what goes bump in our minds. Psychologist Jesse Bering from the University of Arkansas has suggested that both ghosts and God help forge a more honest society by convincing people that they are constantly being watched.[21] Bering and his team tested their idea by carrying out a somewhat strange experiment. In their study, students were asked to complete an intelligence test. The test had been carefully constructed to ensure that the students could cheat if they wanted to, and that the experimenters could secretly monitor each person's level of deception. Just before taking the test, a randomly selected group of the students was told that the test room was apparently haunted. As predicted by the 'ghosts make people more honest' theory, the students who thought that they were in a haunted room were far less likely to cheat on the test.

However, perhaps the most popular theory to account for the evolution of ghostly experiences concerns the 'Hypersensitive Agency Detection Device'.[22] Oxford University psychologist Justin Barrett believes that the idea of 'agency' – being able to figure out why people act the way they do – is essential to our everyday interactions with one another. In fact, it is so important that Barrett thinks the part of the brain responsible for detecting such agency often goes into overdrive, causing people to see human-like behaviour in even the most meaningless stimuli. In the 1940s psychologists Fritz Heider and Mary-Ann Simmel conducted a now classic experiment that provides a beautiful illustration of Barrett's point. Heider and Simmel created a short cartoon animation in which a large triangle, small triangle and a circle moved in and out of a box. They then showed the meaningless cartoon to people and asked them to describe what was happening. Most people instantly created elaborate stories to explain the cartoon, saying, for example, that perhaps the circle was in love with the little triangle, and the big triangle was attempting to steal away the circle, but that the little triangle fought back, and the small triangle and circle eventually lived happily ever after.

In short, people saw agency where none existed. Barrett believes that the same concept helps explain God, ghosts and goblins. According to the theory, many people are very reluctant to think that certain events are meaningless, and are all too eager to assume that they are the work of invisible entities. They might, for instance, experience an amazing stroke of good luck and assume it is angels at work, be struck down with an illness and see it as evidence of demons, or hear a creaking door and attribute it to a ghostly woman in white. If Barrett is right, ghosts are not the result of superstitious thinking. Neither are they spirits returning from the dead. Instead, they are simply the price we pay for having remarkable brains that can effortlessly figure out why other people behave the way they do. As such, ghosts are an essential part of our everyday lives.

6. MIND CONTROL

In which we climb inside the head of the world's greatest thought-reader, discover whether hypnotists can make us act against our will, infiltrate some cults, learn how to avoid being brainwashed and investigate the psychology of persuasion.

Think of any number between one and a hundred. Feel free to change your mind a few times before deciding on your number. Do you have a number in mind? OK, focus on it. I am getting the impression that you are thinking of . . . number 73. Research suggests that around 1 in 50 of you have just dropped the book in amazement. Unfortunately, the same work also shows that the vast majority of you are totally underwhelmed by my thought-reading skills.

However, imagine that I had been able to accurately name the number that you were thinking of. Moreover, imagine that my remarkable telepathic powers were not limited to naming numbers, but also worked with shapes, names, locations, and colours. Finally, imagine that my abilities stretched far beyond rummaging around in the contents of your mind, and that I also had the ability to actually control your behaviour. Over the years a small number of people have claimed to possess these abilities. These rather curious individuals are not interested in staring into a crystal ball, talking with the dead or analysing your astrological chart. Instead, they appear to have an uncanny and remarkable ability to play directly with your mind. How do they appear to achieve the impossible? Do their feats constitute compelling evidence for the paranormal, or is there some subtle and mysterious psychology at work?

To find out, we are going to journey deep into the world of a remarkable telepath, meet a mind-reading horse and spend some time with a terrifying mind control expert. Our journey starts over a hundred years ago, with one of the world's first thought-readers.

Thought-reading on the Brain

Washington Irving Bishop was, by any measure, a remarkable man.[1] Born in 1856 in New York City, Bishop was raised primarily by his mother, Eleanor, who made her living as an actress, opera singer and part-time medium. Eleanor was a colourful character who was frequently at the centre of controversy. In 1867, for example, she attempted to divorce her husband Nathaniel on the grounds that he had tried to murder her. In 1874 Eleanor attended Nathaniel's funeral and, despite the two of them having been separated for the last seven years, was apparently so moved by the event that she felt the need to throw herself on Nathaniel's casket as it was lowered into the grave. A few weeks later she claimed that Nathaniel had been deliberately poisoned by a mysterious enemy, and demanded that his body be exhumed. A thorough examination of the body failed to produce any evidence of foul play.

Bishop didn't excel at college and, perhaps helped by his mother's connections with Spiritualism, ended up working as the manager for a well-known stage medium of the day named Annie Eva Fay. At the start of her act Fay would place a chair and various musical instruments in a large open-fronted cabinet. Next, she would invite several audience members onto the stage, and ask them to tie her to the chair. A curtain would be drawn across the front of the cabinet and Fay would allegedly summon the spirits. After a few moments the spirits would apparently make their presence known by playing the instruments and then throwing them out of the cabinet. Various rumours circulated about how Fay was producing these seemingly miraculous phenomena, with some going as far as to suggest that she smuggled her young son into the cabinet by secreting him under her dress. The truth was far more straightforward. Fay was a skilled escapologist who was able to free herself from the chair, play the instruments, throw them out of the cabinet, and then wriggle back into her bonds.

After a few months Bishop fell out with Fay over a financial matter, and decided to make his own music hall debut by presenting a public exposure of her entire act. Although all went well initially, audiences soon began to tire of hearing about Fay's secrets, and Bishop decided to expand his repertoire by exposing the tricks of the trade being employed by other well-known mediums. For reasons that are still not entirely apparent, Bishop thought that the best way of collecting this new material was to attend séances dressed as a woman. Unfortunately, his subsequent accounts of his transvestite exposés failed to capture the public interest, and he was forced to explore alternative ways of attracting an audience. After much trial and error, he eventually developed a skill that would guarantee him international fame and fortune.

He underwent a complete rebranding. Instead of presenting himself as a music hall entertainer, he adopted the far more sombre style of a scientific lecturer. Out went the sensationalist 'once again I put on a dress and discovered the truth' stories, and in came a pair of pince-nez glasses and academic mutton chop sideburns. Perhaps most importantly of all, instead of focusing on exposing the claims of others, Bishop declared that he himself had developed the most uncanny of abilities. Promoting himself as the 'world's first mind-reader', Bishop proudly announced that he was able to demonstrate telepathy on demand.

He started his performances by playing the mystery card, clearly stating that although his newfound ability was not due to psychic powers or the work of the spirits, he did not have an explanation for what he was about to demonstrate. He would then attempt a series of mind-reading stunts. In a typical performance he handed a pin to a spectator and explained that in a few moments the spectator was to hide the pin anywhere in the auditorium. Another member of the audience was asked to ensure that Bishop didn't see where the pin was being concealed. Bishop and his chaperone then walked offstage and the pin was hidden. When he returned, he grasped the first spectator's wrist and led him manically around the auditorium. Eventually, Bishop narrowed down his search to one small area and finally located the hidden pin.

There were many variants on the procedure. Sometimes, for example, he brought a large directory onstage and asked a spectator to secretly choose a name from it. Bishop then used his alleged telepathic skills to identify the chosen name. In perhaps his most famous stunt, he invited a group of five or six people onstage, explained that he would leave the auditorium, and asked them to mime a murder scene in his absence. One person in the group played the role of the murderer and another the victim. After the audience had witnessed the 'murder', Bishop returned and was blindfolded. He then held the wrist of an audience member and asked them to concentrate on the person who had been 'murdered'. After working his way around the group, he correctly worked out who had been playing the role of the victim. Seconds later Bishop successfully identified the 'murderer'.

His amazing demonstrations proved highly successful, and his reputation quickly spread across Europe and America. Bishop's fame encouraged a handful of imitators, with perhaps the best known being one of his former employees, Stuart Cumberland. The level of success enjoyed by the likes of Bishop and Cumberland was reflected in their high society audiences (Cumberland was invited to the House of Commons to read the mind of William Gladstone, later describing the Prime Minister's 'remarkable magnetic influence' in his book *People I Have Read*), as well as their being satirized in well-known comic songs of the period, such as the ever-popular 'Thought-reading on the Brain':

Oh, Mr Cumberland and Irving Bishop too
With the pins you find I'd like to run you through
For you have marr'd my happiness and it is very plain
That all the family now have got thought-reading on the brain

Unfortunately, Bishop's success was short-lived. In 1889, the world-famous mind-reader found himself performing at the Lambs Club in New York City. After successfully completing his 'identify the murderer' and 'find the name in the directory' stunts, he fell to the ground exhausted. He regained consciousness a few moments later and was taken to a bed in the club. Ever the professional, Bishop insisted on performing another feat. The club ledger was duly brought into the bedroom and a name chosen at random. Clearly struggling, he eventually managed to locate the correct name. Immediately after performing what was to be his final stunt, he collapsed back into his bed.

Two doctors were summoned and kept a watchful eye over him throughout the night. In the middle of the following day Bishop, aged just 33, was pronounced dead. The news was quickly conveyed to Bishop's wife in Philadelphia, who promptly made her way to New York City and tracked down her husband's body in a funeral parlour. She was horrified to discover that at some point in the afternoon, and less than 24 hours after his death, her husband had been subjected to an unauthorized autopsy.

Throughout his life Bishop had been prone to cataleptic fits. During these episodes his entire body would become rigid, his breathing very shallow and his heartbeat so slow as to be imperceptible. Because of this, he always carried a card explaining that he might lapse into a cataleptic state, and that no autopsy should be performed until at least 48 hours after his alleged death. At one point he had told a friend that when he was in a cataleptic state he was fully aware of everything that was happening around him, raising the terrifying notion that he was conscious throughout his autopsy.

Why was the autopsy performed so quickly? Throughout his career Bishop boasted of having an exceptional brain. Many historians now believe that this claim may have contributed to his demise, encouraging physicians to carry out a quick autopsy in order to be the first to examine it. Whatever the truth, the autopsy proved a wasted effort. Bishop's brain weighed only slightly more than average and didn't appear at all exceptional.

His mother Eleanor demanded a coroner's inquest, and the doctors who had conducted the autopsy were arrested. However, a jury found in favour of the doctors and the charges against them were dropped. Eleanor remained unconvinced, and made her feelings known by having her son's gravestone read, 'Born May 4th, 1856 – Murdered May 13th, 1889' and publishing a small book describing 'the butchery of the late Sir Washington Irving Bishop'. Eleanor's behaviour became increasingly erratic, and when she passed away in 1918, the famous magician Harry Houdini discovered that she had left him an imaginary estate totalling 30 million dollars.

So how did Bishop achieve his mind reading feats? Did he really possess genuine telepathic powers?

In the early 1880s, Bishop was investigated by a team of well-respected scientists that included the Queen's personal physician, the editor of the *British Medical Journal*, and the famous eugenicist Francis Galton. During the first part of the investigation Bishop successfully performed several stunts, including correctly identifying a selected spot on a table and finding an object that had been hidden on a chandelier. As usual, throughout all of the demonstrations he asked to be in physical contact with an individual who knew the correct answer. Bishop would hold the helper's wrist, or the helper would grasp one end of a walking stick while he held the other. The scientists speculated that Bishop had trained himself to detect the tiny 'ideomotor' movements that had originally been uncovered by Michael Faraday during his investigation into table-tipping. When performing his stunts, Bishop would push and pull his helper in various directions, and the scientists believed that he used tiny changes in resistance to figure out the location of a hidden object, or which member of a group had taken on the role of 'murderer'. The team carried out a second set of trials to discover if they were right. This time, Bishop was asked to try to find a hidden object when his helper was blindfolded and had lost his bearings. He failed. In another trial the walking stick was replaced with a slack watch chain thatprevented any unconscious signals being transferred to Bishop. Once again, he failed. Galton and his fellow scientists concluded that Bishop did possess a remarkable skill, but was not a genuine telepath.

A few years later another amazing mind-reader hit the headlines. However, this time the claim was even more startling because it appeared to provide incontrovertible evidence of animal to human communication.

HOW TO READ MINDS

It is time to get in touch with your inner Bishop. Muscle-reading isn't easy, but there are several simple exercises that will help you develop this remarkable skill.

1. Ask someone to hold out their hand palm up in front of them with their fingers spread apart, and then ask them to concentrate on one of their fingers. Next, lightly push down on each of their fingers with your forefinger. The finger on which they are concentrating will be the one offering the greatest resistance.

2. Arrange four objects in a row on a table, allowing about four inches between each object. Ask someone to stand on your right-hand side and think of one of the objects. Next, take hold of their left wrist with your right hand, placing your fingers on the top of their wrist and your thumb on the bottom of their wrist. Explain that you are going to move their left hand over each of the objects. Ask your guinea pig not to consciously move their left hand but instead to relax their arm and simply 'will' their left hand to move in the correct direction. If you are over the wrong object they should think of the phrase 'move on', whereas if you are over the correct object then they should think of the word 'stop'. Now move their left hand over each of the objects and try to discover the chosen object by feeling when you encounter most resistance to movement.

3. Time for a full test of muscle-reading. Ask your volunteer to go into a room and hide a small object. Next, hold their wrist as instructed above. Take the weight of their right arm and keep them close by your side. Ask them not to focus on the location of the object, but rather on the direction that you have to proceed in order to move towards it. Stand in the centre of the room and take a step forward. If there is a feeling of resistance then go back to the centre of the room and head in another direction. Keep on doing this until you feel the least resistance. When you think that you are near to the object, have your helper imagine a straight line between their hand and the object. When you feel the hand move in that direction, follow along the line and you should be able to find the object.

As muscle reading is tricky to master, some mind-readers perform the following trick to develop their skills without worrying about the risk of failure.

Before performing the demonstration, find a deck of cards, separate the red and black cards, and place the stack of red cards on top of the stack of black ones.

Next, find a willing spectator, fan out the top section of the deck (containing only red cards) face down between your hands, and ask your guinea pig to remove a card. Ask them to look at the card but keep its identity secret. While this is happening, close up the deck, and then spread out the bottom section of the deck face down between your hands. Now the spread contains only black cards, whereas previously the participant was choosing from only red cards.

Ask the spectator to replace their card face down into the spread, and close up the deck. Their card will now be the only red card in the black section. Explain that you will try to guess the identity of their card. As you say this, turn the deck towards you, and quickly spread the pack between your hands. You will easily see your spectator's chosen card because it will be the only red card in the black section.

Now shuffle the deck and spread it face up on the table. Hold onto your spectator's wrist as before and lead them along the cards. See if you can pick up subtle cues from their hand. Slowly home in on the section of the deck with their card in it and then, with a dramatic flourish, announce the name of the card.

Straight From the Horse's Mouth

Wilhelm von Osten was the most curious of men.[2] Born in 1834, this unassuming German mathematics teacher had a passion for odd ideas. A strong advocate for the then relatively new theory of evolution, von Osten believed that animals were just as bright as humans, and that the world would be a better place if people could communicate with other species and appreciate their amazing intellect. In 1888, von Osten retired from teaching, moved to Berlin and spent the remainder of his life pursuing his dream.

His initial attempts at uncovering the hidden genius of the animal kingdom involved trying to teach the fundamentals of mathematics to a cat, a bear and a horse. Each day, von Osten would draw numbers on a blackboard and encourage his class to count by moving their paws or hooves an appropriate number of times. In what must be one of the most bizarre school reports ever written, he later described how the cat quickly lost interest in the enterprise and the bear was downright hostile. The horse, however, proved an attentive student and quickly learned how to stamp out any number written on the blackboard. Flushed by this initial success, von Osten expelled the cat and bear from his classroom, and focused solely on equine pupils.

Von Osten acquired a Russian trotting horse called Hans, and together the two of them embarked on another four years of daily training in the fundamentals of mathematics.

In 1904, the duo felt ready for their first public demonstration. A small crowd of spectators were invited into von
Osten's courtyard and asked to form a semi-circle around
'Clever' Hans. Von Osten, sporting a long white beard, loose-
fitting smock, and floppy black hat, stood to the side of the
animal while members of the audience called out mathematical problems. Each time, Clever Hans indicated his answer by
stamping his hoof against the cobbles. It was an impressive
performance, with Hans correctly answering simple addition
and subtraction problems, as well as more complex sums with
fractions and square roots. Encouraged by this initial success,
von Osten worked with Hans to increase his repertoire. Over
time he taught the horse to tell the time, choose which musical
tones would improve a harmony, and even answer questions
by nodding or shaking his head.

In 1904, psychologist Oskar Pfungst decided to investigate Clever Hans, unaware that the work would guarantee
him a place in almost every psychology textbook for the next
hundred years. During Pfungst's carefully controlled studies
members of the public were then asked to present Hans with
pre-planned questions. To ensure a well-motivated participant, Pfungst rewarded Clever Hans with a small piece of
bread, carrot or sugar each time he responded (interestingly,
this same procedure still works well with most undergraduate
students today). It was not all easy going. Both von Osten and
Clever Hans were prone to rage, and Pfungst received several
bites during the investigation, the majority of which came
from the horse. Regardless, the young German researcher
methodically worked his way through a series of ground-
breaking tests.

In one study, a series of number cards were first oriented

in such a way as to ensure that Clever Hans, von Osten and a questioner could all see the front of the cards. A question was then asked, and Clever Hans stamped his hoof to indicate which card contained the answer. Under these circumstances, Clever Hans demonstrated an impressive 98 per cent success rate. However, when Pfungst altered the orientation of the cards to ensure that only Clever Hans could see the faces of the cards his hit rate dropped to an unimpressive 6 per cent. In another test, von Osten whispered two numbers into Hans's ear and asked him to add them up. Time and again, Hans stamped out the right response. However, when von Osten whispered one number and Pfungst another, with neither man knowing the other's number, Hans failed to produce the correct answer.

Pfungst obtained the same pattern in test after test. Whenever von Osten or a questioner knew how Clever Hans should respond, the horse did well. When no one knew the correct response, Hans failed. Pfungst concluded that Clever Hans was not thinking for himself but rather responding to involuntary signals in the facial expressions and body language of those around him. For years von Osten had not been talking to the animals, but instead chatting to himself.

Researchers across the world quickly realized that the general principle uncovered by Pfungst, namely that experimenters may be unknowingly persuading participants to act in a desired way, could have major implications for their work.

Scientists went in search of the phenomenon – dubbed the 'Clever Hans effect' – and found it in several different settings. In one classic experiment rats were randomly divided into two groups, and then given to students who were told that the groups had been selectively bred for good and poor performance in navigating mazes.[3] In fact, there was no special breeding at all. The students then ran the rats through mazes and reported results in line with their expectations, with the allegedly 'bright' rats making 51 per cent more correct responses than the allegedly 'dull' rats.

Similarly, in research called the 'Pygmalion experiment', Harvard psychologist Robert Rosenthal administered a test to an entire year-group of children, telling their teachers that it represented a new technique for predicting intellectual 'blooming'.[4] Teachers were then led to believe that they had been given the names of the children in their class who had obtained the highest scores. In reality, Rosenthal's test was an ordinary measure of intelligence, and the names of the alleged 'bloomers' were chosen at random. At the end of the school year, the children were given the same intelligence test, and the children randomly identified as intellectual 'bloomers' scored an average of 15 points more than the other children.

According to Gary Wells, of Iowa State University, this theory could even lead to police officers unwittingly biasing witnesses to choose certain suspects from line-ups, by using exactly the same type of unconscious nonverbal signal that influenced Clever Hans over a hundred years ago.[5]

This work made researchers recognize the need to guard against the Clever Hans effect by hiding certain aspects of a study from both the participants and experimenters. 'Blind' methods are now the gold standard of good science. And all because of a mathematical horse.

Both Bishop and Clever Hans appeared to be able to read people's thoughts. In reality, both were simply responding to the involuntary signals given out by those around them.

Other mind wizards have focused more on trying to control those thoughts and so persuade people to behave in certain ways. But is it really possible to take over someone's mind and manipulate them like a puppet? Over the years several novelists and filmmakers have suggested it is, but what is the fact behind the fiction? Can someone be hypnotized to act against their will?

The Svengali Effect

In 1894 George du Maurier published his classic novel *Trilby*. The plot features a rogue hypnotist named Svengali, who places heroine Trilby O'Ferrall into a deep trance and then exploits her for his own benefit. In addition to being the second-bestselling novel of its day (outperformed only by Bram Stoker's *Dracula*), and giving rise to the Trilby hat, du Maurier's novel encouraged the public to believe that some people have the power to make others act against their will. But is this really the case?

Around the turn of the last century several researchers tackled the issue by placing people in trance states and asking them to carry out various questionable acts, such as committing a mock murder or throwing a glass of 'acid' (actually water) in the face of the experimenter.[6] Although many of the participants did stab others with rubber daggers and soak researchers, the work was not carried out under well controlled conditions and so generated more questions than answers. In the mid-1960s, University of Pennsylvania psychologists Martin Orne and Fredrick Evans decided to take a more rigorous look at the issue.[7]

Orne found some highly suggestible students and tested them one at a time. Each was put into a trance and then asked to sit before an open-fronted box. A researcher placed a harmless green tree snake in the box and the participants were told that they had an irresistible urge to pick up the snake. All of them went along with the suggestion and removed the snake. Next, the experimenters put on a pair of long thick gloves and brought forth a genuinely dangerous red-bellied black snake. They explained that this was one of the most venomous snakes in the world and could kill a human with a single bite. The snake was placed into the box and all of the participants were told that they had an irresistible urge to pick it up. Amazingly, all of them tried to carry out the action, and it was only as they placed their hands into the box that they discovered the researchers had secretly slid a glass plate in front of the snake.

On the face of it, Orne and Evans appeared to have persuaded the hypnotized students to act against their best interests. However, a second stage of the study was cleverly designed to discover if that was really true. The experimenters found a group of six highly non-suggestible students, didn't bother trying to put them into a trance, and instead simply asked them to pretend to be hypnotized. Surprisingly, all of them were also willing to try to pick up both the harmless snake and its highly venomous counterpart. It was clear that the results obtained in the first stage of the study were not due to hypnosis. To discover why the students were willing to risk their lives during the experiment, the researchers then asked their non-suggestible participants what they were thinking when they reached for the poisonous snake. Nearly all of them explained that they knew they were taking part in a study and so were convinced that the experimenter wouldn't let them come to any harm. These findings suggested that it isn't possible for researchers to properly evaluate whether people can be made to act against their will when hypnotized. University ethics committees wouldn't allow participants to be put into a situation that was genuinely risky and, even if they did, participants might carry out a dangerous act simply because they believed that they were safe.

However, when researchers took a careful look back at older investigations into the alleged Svengali effect they discovered one demonstration that overcame this problem. Around the turn of the last century, hypnotist and researcher Jules Liegeois conducted a rather unusual demonstration during a conference held at the Salpêtriére School in Paris. Liegeois placed a young woman into a trance, handed her a rubber knife, said that it was a genuine knife, and asked her to stab someone in the audience. The woman promptly obliged. Unfortunately, Liegeois didn't think to ask someone who wasn't hypnotized to carry out the same test, and so incorrectly concluded that the demonstration showed that those in hypnotic trances could be made to behave in a way that was not in their best interests. However, once most of the conference-goers had left the room, a group of mischievous medical students told the still-hypnotized woman that she should remove her clothing. The woman would have realized that whereas stabbing someone with a rubber dagger was all good fun, complying with this suggestion was going to be genuinely embarrassing. She didn't strip off. In fact, she stood up and ran out of the room. Interestingly, there has been only one attempt to replicate this fascinating, but unethical, study. In the 1960s a University researcher randomly selected a young female volunteer, sat her in front of a group and suggested that she remove her clothes. The professor was horrified to discover his volunteer rapidly starting to unbutton her clothing and quickly called a halt to the demonstration. It was only later that he discovered that he happened to have chosen a professional stripper as his subject.

HOW TO HYPNOTIZE A CHICKEN

Ormond McGill was a talented stage hypnotist. Born in 1913, he worked under the stage name of 'Dr Zomb', and pioneered many of the techniques used by modern-day performers. McGill's 1947 book, *The Encyclopedia Of Genuine Stage Hypnotism*, describes how chickens can be positioned to ensure that they become motionless and appear hypnotized. According to McGill, all you need to do is carefully catch the bird by its neck, place it on its front on a table, and rest its head horizontally. Finally, draw a two-foot-long chalk line on the table, directly out from its beak. The chicken will then lie motionless on the table (see photograph).

While hypnotized, the chicken can be made to eat an onion, wear X-ray glasses, and perform a striptease. Just kidding. Actually, rather than being hypnotized, the lack of movement is due to tonic immobility, wherein the chicken is engaging in a defensive mechanism intended to put off potential predators by feigning death. To appear to awaken the animal from the deep trance, simply push the chicken's head away from the chalk line.

Despite the mass of films and books suggesting otherwise, the scientific evidence suggests that it is not possible to make people act against their will by hypnotizing them. However, work into other forms of mind control has yielded far more positive and worrying results. To find out more we have to explore the dark and murky world of cults.

From Monkey Salesman to Charismatic Preacher

Born in 1931, Jim Jones grew up in a rural Indiana community.[8] Later described by some of his neighbours as a 'really weird kid', Jones spent much of his childhood exploring religion, torturing animals, and discussing death. He also exhibited an early interest in preaching, with one childhood friend recalling how Jones once draped an old sheet over his shoulders, formed a group of other children into a makeshift congregation, and promptly gave a sermon pretending to be the Devil. In his teens he enrolled as a student pastor in a local Methodist Church, but left when the church leaders banned him from preaching to a racially mixed congregation. In 1955, aged just 24, Jones rounded up a small flock of faithful followers and founded his own church, the Peoples Temple. Rather bizarrely, he funded this ambitious venture by going door-to-door selling pet monkeys. When he wasn't engaged in monkey business he spent time honing his public speaking skills and soon built a considerable reputation as a highly charismatic preacher.

Jones' initial message was one of equality and racial integration. Practising what he preached, he encouraged his followers to help provide food and employment for the poor. Word of his good deeds soon spread, resulting in almost a thousand people flocking to his church. Jones continued to use his influence to help enrich the community, opening both a soup kitchen and a nursing home. In 1965 he claimed to have had a vision that the Midwest of America would soon be the target for a nuclear strike, and persuaded about a hundred members of his congregation to follow him to Redwood Valley in California. He still focused on supporting those most in need, helping drug addicts, alcoholics and the poor.

By the early 1970s storm clouds were gathering. He was asking for a greater level of commitment from his followers, urging them to spend holidays with other Temple members rather than their families, and give their money and material possessions to the church. In addition, Jones had developed a serious drug habit and had become increasingly paranoid about the idea of the American government trying to destroy his church. Local journalists eventually started to take an interest in the stories of unhealthy levels of commitment emerging from the Peoples Temple, causing Jones to attempt to escape unwanted scrutiny by shifting his headquarters to San Francisco. Here his preaching again proved highly successful, and within a few years the Temple congregation had doubled in size. However, before long journalists again started to write articles that criticized him, prompting him to decide to leave America and build his own 'utopian' community abroad.

He carefully considered several countries before deciding to set up his self-supporting commune in Guyana on the northern coast of South America. From Jones' perspective it was a wise choice, in part, because Guyanese officials could be easily bribed, allowing him to receive illegal shipments of firearms and drugs. In 1974 he negotiated a lease on almost 4,000 acres of remote jungle in the north-west of the country. Modestly naming the plot 'Jonestown', the charismatic preacher and several hundred of his followers packed their bags and moved to Guyana. It was a harsh existence. Jonestown was isolated, suffered from poor quality soil, and the nearest water supply could be reached only after a seven-mile hike along muddy roads. Severe diarrhoea and high fevers were common. In addition to working 11-hour days, Temple members were also expected to attend long evening sermons and classes in socialism. A variety of punishments were administered to those who neglected their duties, including imprisonment in a small coffin-shaped wooden box, and being forced to spend hours at the bottom of a disused well.

On the 17 November 1978, US Congressman Leo Ryan travelled to Guyana to investigate rumours of people being held at Jonestown against their will. When he arrived, Ryan initially heard nothing but praise for the new community. However, towards the end of the first day of his visit a small number of families secretly informed Ryan that they were far from happy and eager to leave. Early the following morning 11 Temple members sensed a growing feeling of danger and desperation in Jonestown, and secretly made their escape by walking 30 miles through the surrounding dense jungle. Later that day Ryan and a small number of defectors headed to a nearby airstrip and attempted to board planes to return to America. Armed members of the Temple's 'Red Brigade' security squad opened fire, killing Ryan and several members of his group. Ryan became the only Congressman in the history of the US to be murdered in the line of duty.

Sensing his world crumbling around him, Jones gathered together the residents of Jonestown, told them that Ryan and his party had been killed, explained that the American government would now seek revenge on the community, and urged everyone to participate in a mass act of 'revolutionary suicide'. Large drums of grape-flavoured juice laced with cyanide were brought out, and Jones ordered everyone to drink the liquid. Parents were urged to first administer the poison to their children and then drink it themselves. An audiotape made at the time shows that whenever followers were reluctant to participate, Jones urged them to join in, proclaiming 'I don't care how many screams you hear, I don't care how many anguished cries, death is a million times preferable to this life. If you knew what was ahead of you, you'd be glad to be stepping over tonight.' Over 900 people died during the ritual, including around 270 children. Although several armed Temple guards had surrounded the group, it appears that the majority of the followers willingly killed themselves, with one woman writing 'Jim Jones is the only one' on her arm during the episode. Up until 11 September 2001 the deaths represented the greatest single loss of American civilian life in a non-natural disaster.

For over 30 years psychologists have speculated as to how Jim Jones persuaded so many people to take their own lives, and parents to murder their children. Some have pointed out that the majority of the Temple congregation were psychologically vulnerable individuals desperate to believe Jones' message of equality and racial harmony. Jones referred to Jonestown as the 'promised land' and described it as a place where parents could raise their children away from the racial abuse that had scarred their own lives. His mission was also attractive because it provided people with a strong sense of purpose, a relief from feelings of worthlessness, and made them part of a large family of caring and like-minded individuals. As one survivor memorably put it, 'Nobody joins a cult . . . you join a religious organization or a political movement, and you join with people you really like.' Although these factors clearly played a part in the Jonestown tragedy, they are far from the full picture. People are often attracted to religious and political organizations because they offer a sense of purpose and extended family, but most would be unwilling to lay down their lives for the cause. Instead, psychologists believe that Jones' influence relied on four key factors.

First, Jones was skilled at getting his foot in the door.

Getting a Foot in the Door

In a now classic study carried out by Jonathan Freedman and Scott Fraser of Stanford University, researchers posed as volunteer workers and went from door-to-door explaining that there was a high level of traffic accidents in the area and asking people if they would mind placing a sign saying 'DRIVE CAREFULLY' in their gardens.[9] This was a significant request because the sign was very big and so would ruin the appearance of the person's house and garden. Perhaps not surprisingly, few residents agreed to display it. In the next stage of the experiment, the researchers approached a second set of residents and asked them to place a sign saying 'BE A SAFE DRIVER' in their garden. This time the sign was just three inches square, and almost everyone accepted. Two weeks later, the researchers returned and now asked the second set of residents to display the much larger sign. Amazingly, over three-quarters of people agreed to place the big ugly placard. This concept, known as the 'foot in the door' technique, involves getting people to agree to a large request by first getting them to agree to a far more modest one.

Jones used the technique to manipulate his congregation. Followers would first be asked to donate a small amount of their income to the Temple, but over time the amount required would rise until they had given all of their property and savings to Jones. The same applied to acts of devotion. When they first joined the church, members were asked to spend just a few hours each week working for the community. As time passed, these few hours expanded little by little until members were attending long services, helping to attract others into the organization and writing letters to politicians and the media. By ratcheting up his requests slowly, Jones was using the 'foot in the door' technique to prepare his followers to make the ultimate sacrifice. But this technique is only successful if people do not draw a line in the sand and speak out against the increased demands. The second psychological technique employed by Jones was designed to quell this potential rebellion.

All Together Now

In the 1950s, American psychologist Solomon Asch conducted a series of experiments into the power of conformity.[10] Participants were asked to arrive at Asch's laboratory one at a time and were introduced to about six other volunteers. Unbeknownst to each participant, all of these other volunteers were actually stooges who were working for Asch. The group, made up of the participant and stooges, were sat around a table and told that they were about to take part in a 'vision test'. They were then shown two cards. The first card had a single line on it, while the second card contained three lines of very different length, one of which was the same

length as the line of the first card. The group were asked
to say which of the three lines on the second card matched
the line on the first card. They had been seated in such a
way as to ensure that the genuine participant answered last.
Everyone was asked to voice their answers and each of the
'volunteers' always gave the same one. For the first two trials,
all of the stooges gave the correct response to comparing the
lines, while on the third trial the stooges all gave an incorrect
answer. Asch wanted to discover what percentage of participants would conform to peer pressure and give an obviously
incorrect answer in order to go along with the group. Amazingly, 75 per cent of people conformed. In a slight variation
on the procedure, Asch had just one of the stooges break with
the group and give a different answer. This one dissenting
voice reduced the amount of conformity to around 20 per
cent.

The Peoples Temple was a huge experiment in the psychology of conformity. Jones was aware that any dissent would
encourage others to speak out and so tolerated no criticism.
To help enforce this regime, Jones had informers befriend
those thought to be harbouring doubts about the Temple,
with any evidence of dissent resulting in brutal beatings or
public humiliation. He also split up any groups that were
likely to share their concerns with each other. Families were
separated, with children first being seated away from their
parents during services and later placed into the full-time care
of another church member. Spouses were encouraged to participate in extramarital sexual relationships to loosen marital
bonds. Similarly, the dense jungle around Jonestown ensured
that the community was completely cut off from the outside
world and had no way of hearing any dissenting voices from

those not involved. The powerful and terrible effects of this intolerance of dissent emerged during the mass suicide. An audiotape of the tragedy revealed that at one point a woman openly declared that the babies deserved to live. Jones acted quickly to quell the criticism, stating that babies are even more deserving of peace and that 'the best testimony we can give is to leave this goddamn world'. The crowd applaud Jones, with one man shouting 'It's over, sister . . . We've made a beautiful day', and another adding, 'If you tell us we have to give our lives now, we're ready'.

But Jones was not just concerned with getting his foot in the door and quashing any dissent. He also employed a third psychological weapon to help control the minds of his followers – he appeared to have a hotline to God and be able to perform miracles.

Wonder of Wonders, Miracle of Miracles

Many people followed Jones because he appeared to be able to perform miracles. During services Jones would ask those suffering from any illnesses to make their way to the front of the church. Reaching into their mouths, he would dramatically pull out a horrid mass of 'cancerous' tissue and announce that they were now cured. Sometimes the lame would apparently be instantly healed, with Jones telling them to throw away their walking aids and dance back up the aisle. He also claimed to hear the voice of God, calling out to people in the congregation and accurately revealing information about their lives. On one occasion more people than expected turned up for a service and Jones announced that he would feed the multitude by magically producing more food. A few minutes later, the door swung open and in walked a church member carrying two large trays filled with fried chicken.

It was all a sham. The 'cancers' were actually rancid chicken gizzards that Jones concealed in his hand prior to 'pulling' them from people's mouths. The curing of the 'lame' was created by a small inner circle of highly devoted followers pretending that they couldn't walk. The information about the congregation was not God-given, but instead obtained by members of Jones' 'inner circle' sifting through people's rubbish bins for letters and other useful documentation. These individuals later described how they willingly assisted Jones because he told them that he was conserving his genuine supernatural powers for more important matters. And the miracle of the deep fried chicken? One member of the congregation later described how he saw the bearer of the trays arrive at the church a few moments before the miracle, armed with several buckets of food from Kentucky Fried Chicken. When Jones found out about the comment he put a mild poison in a piece of cake, gave it to the dissenting church member, and announced that God would punish his lies by giving him vomiting and diarrhoea.

So was Jones' mind control just about getting his foot in the door, creating conformity and performing miracles? In fact, there was also the important issue of self-justification.

On Behaviour and Belief

In 1959 Stanford University psychologist Elliot Aronson conducted a revealing study into the relationship between belief and behaviour.[11] Let's turn back the hands of time and imagine that you are a volunteer in that experiment.

When you arrive at Aronson's laboratory, a researcher asks you whether you would mind participating in a group discussion about the psychology of sex. Drooling, you say that you are open to the idea. The researcher then explains that some people have become very self-conscious during the discussion and so now all potential volunteers have to pass an 'embarrassment' test. You are handed a long list of highly evocative words (including many containing four letters) and two passages containing vivid descriptions of sexual activity. The researcher asks you to read both the list and passages out loud, while he rates the degree to which you are blushing. After much sanctioned cursing, the researcher says that the good news is that you have passed the test and so can now take part in the group discussion. However, the bad news is that the 'embarrassment' test has taken longer than anticipated, so the discussion has already started and you will just have to listen to the group this time around. The researcher shows you into a small cubicle, explains that all of the group members sit in separate rooms to ensure anonymity, and asks you to wear some headphones. You don the headphones and are rather disappointed to discover that after all you have been through, the group is having a rather dull discussion about a book called *Sexual Behavior in Animals*. Finally, the researcher returns and asks you to rate the degree to which you want to join the group.

Like many psychology experiments, Aronson's study involved a considerable amount of deception. In reality, the entire experiment was not about the psychology of sex, but the psychology of belief. When participants arrived at the laboratory they were randomly assigned to one of two groups. Half of them went through the procedure described above, and were asked to read out highly evocative word lists and graphic passages. Those in the other group were asked to read out far less emotionally charged words (think 'prostitute' and 'virgin'). Everyone then heard the same recorded group discussion and was asked to rate the degree to which they valued being a member of the group. Most psychologists in Aronson's day would have predicted that those who underwent the more embarrassing procedure would end up liking the group less because they would associate it with a highly negative experience. However, Aronson's work into the psychology of self-justification had led him to expect a quite different set of results. Aronson speculated that those who had read out the more evocative sexual material would justify their increased embarrassment by convincing themselves that the group was worth joining, and end up thinking more highly of it. Aronson's predictions proved correct. Even though everyone had heard the same recording of the group discussion, those who underwent the more extreme embarrassment test rated joining the group as far more desirable than those in the 'prostitute and virgin' group.

Aronson's findings help explain why many groups demand that potential members undergo painful and humiliating initiation rituals. American college fraternities make freshmen eat unpleasant substances or strip naked, the military put new recruits through extreme training, and medical interns are expected to work night and day before becoming fully fledged doctors. Jones used the same tactics to encourage people to feel committed to the Peoples Temple. Members of the congregation had to endure long meetings, write self-incriminating letters, give their property to the Temple, and allow their children to be raised by other families. If Jones suspected someone of behaving in a way that was not in the interests of the Temple, he would ask other members of the congregation to punish them. Common sense would predict that these acts would drive people away from both Jones and the Peoples Temple. In reality, the psychology of self-justification ensured that it actually moved them closer to the cause.

The mind control exhibited by the likes of Jim Jones does not involve any hypnotic trances or prey on the suggestible. Instead, it uses four key principles. The first involves a slow ratcheting up of involvement. Once a cult leader has his foot in the door, they ask for greater levels of involvement until suddenly followers find themselves fully immersed in the movement. Second, any dissenting voices are removed from the group. Sceptics are driven away and the group is increasingly isolated from the outside world. Then there are the miracles. By appearing to perform the impossible, cult leaders often convince their followers that they have direct access to God and therefore should not be questioned. Finally, there is self-justification. You might imagine that asking someone to carry out a bizarre or painful ritual would encourage them to dislike the group. In reality, the opposite is true. By taking part in these rituals followers justify their suffering by adopting more positive attitudes towards the group.

Of course, it would be nice to think that if the group had not been so isolated from society, it might have been possible to undo the effects of these techniques, explain the madness of their ways, and avert a major tragedy. However, our final sojourn into the world of cults suggests that this is a naive view of those that have fallen under the spell of a charismatic leader.

HOW TO AVOID BEING BRAINWASHED

It is easy to avoid having your mind controlled providing that you look out for the following four danger signs.

1. Do you feel as if the 'foot in the door' technique might be at work? Did the organization or person start by asking you to carry out small acts of commitment or devotion, and then slowly increase their requirements? If so, do you really want to go along with their requests or are you being manipulated?

2. Be wary of any organization that attempts to distance you from a dissenting point of view. Are they trying to cut you off from friends and family? Within the organization, is dissent and open discussion squashed? If the answer to either of these questions is 'yes', think carefully about any involvement.

3. Does the leader of the organization claim to be able to achieve paranormal miracles? Perhaps healings or acts of prophecy? However impressive, these are likely to be the result of self-delusion or deception. Don't be swayed by supernatural phenomena until you have investigated them yourself.

4. Does the organization require any painful, difficult or humiliating initiation rituals? Remember that these may well be designed to manufacture an increased sense of group allegiance. Ask yourself whether any suffering is really needed.

The End of the World is Nigh

In the early 1950s, psychologist Leon Festinger came across an item in a local newspaper describing how a cult-like group was predicting the end of the world. According to the article, a woman named Marian Keech was indulging in a spot of automatic writing and claimed that the messages were from aliens. Keech had convinced a small group of 11 followers that there would be a great flood on 21 December 1954, but that they shouldn't worry because a flying saucer would rescue them just before the disaster.

Festinger wondered what would happen to Keech and her followers when the anticipated flood and flying saucers failed to materialize. To find out, he secretly had several undercover observers infiltrate the group and carefully record every psychological twist and turn. Describing his findings in a book entitled *When Prophecy Fails* (which gives you a clue as to whether the spacecraft actually arrived), Festinger produced a fascinating insight into the psychology of the cult.[12]

A few days before the group expected the world to end, Mrs Keech and her followers were buoyant, with one member even baking a large cake depicting the mother ship and bearing the iced message 'Up in the air!' On the big day the group were nervous and excited. The aliens had sent several messages to Keech explaining that they would knock on her door at midnight and lead the group to their nearby flying saucer (apparently there was no parking directly outside the house).

The aliens had also said that it was vital that no one had any metal on them, and so for several hours before the anticipated visit the group members replaced their belts with string, carefully cut any zippers from their clothing, and ripped out the eyelets from their shoes. Keech's books of automatic scribbling were then placed in a large shopping bag and everyone waited for the aliens.

Just after midnight it became obvious that the extraterrestrial visitors were a no-show. The group sat in stunned silence and spent the next four hours trying to find an explanation for what had happened. When they failed, Keech began to cry. However, a few hours later she said that she had received another message from the aliens, explaining that the predicted cataclysm had been called off because the group had managed to spread light upon the world. Festinger's study illustrates how people have a remarkable ability to explain away evidence rather than change their cherished beliefs. This 'I have made up my mind, don't confuse me with the facts' approach helps their beliefs emerge unscathed through even the most devastating attacks. Only two members of Keech's group, both of whom were lightly committed to begin with, abandoned their belief in the guru's writings.

Festinger noticed that rather than walk away with their tails between their legs, many members of the group subsequently became especially eager to spread the word. Prior to the failed prediction the group shunned publicity and only gave interviews grudgingly. Immediately afterwards they contacted the media and began an urgent campaign to spread their message. Festinger explained this curious behaviour by speculating that they were trying to convince themselves that their belief was correct by convincing others, feeling that if lots of people believe in something then clearly there must be something in it.

Eventually the group broke up and everyone went their separate ways. Some took to the road, travelling from one flying saucer convention to the next spreading the good word. Others returned to their previous lives. Keech became increasingly concerned about attention from law enforcement agencies and went into hiding. After spending several years in Peru, Keech returned to Arizona and continued to claim to be in contact with aliens until her death in 1992.

It would be comforting to think that the type of mind control discussed in this chapter is limited to the somewhat bizarre and esoteric world of cults. Comforting, but wrong. In fact, you frequently encounter exactly the same principles of persuasion in everyday life. Salespeople use the 'foot in the door' technique to secure a sale. Politicians attempt to silence dissenting voices and misdirect you away from information that they don't want you to see. Marketeers make liberal use of the 'self-justification' principle, well aware that the more you pay for a product, the more mental hoops you will jump through to justify the purchase. And advertising agencies know that, in the same way that Marian Keech's followers boosted their own beliefs by trying to convert others, you recommend products to friends and colleagues in an attempt to convince yourself that you made the right decision. Although the contexts in which the principles operate differ, the psychology is exactly the same. The practitioners of mind control are not restricted to cult leaders and religious sects. Instead, they walk among us on a daily basis.

7. PROPHECY

In which we find out whether Abraham Lincoln
really did foresee his own death, learn how to
control our dreams and delve deep into the
remarkable world of sleep science.

Aberfan is a small village in South Wales. In the 1960s, many of those living there worked at a nearby colliery that had been built to exploit the large amount of high quality coal in the area. Although some of the waste from the mining operation had been stored underground, much of it had been piled on the steep hillsides surrounding the village. Throughout October 1966 heavy rain lashed down on the area and seeped into the porous sandstone of the hills. Unfortunately, no one realized that the water was then flowing into several hidden springs and slowly transforming the pit waste into soft slurry.

Just after nine o'clock on the morning of 21 October, the side of the hill subsided and half a million tonnes of debris started to move rapidly towards the village. Although some of the material came to a halt on the lower parts of the hill, much of it slid into Aberfan and smashed into the village school. Several classrooms were instantly filled with a ten metre deep mass of slurry. The pupils had left the school assembly hall a few moments before, having sung the hymn 'All things bright and beautiful', and so were just arriving in their classrooms when the landslide hit. Parents and police rushed to the school and frantically began digging through the rubble. Although a handful of children were pulled out alive during the first hour or so of the rescue effort, no other survivors emerged. One hundred and thirty-nine schoolchildren and five teachers lost their lives in the tragedy.

Psychiatrist John Barker visited the village the day after the landslide.[1] Barker had a longstanding interest in the paranormal and wondered whether the extreme nature of events in Aberfan might have caused large numbers of people to experience a premonition about the tragedy. To find out, Barker arranged for the *Evening Standard* newspaper to ask any readers who thought they had foreseen the Aberfan disaster to get in touch. He received 60 letters from across England and Wales, with over half of the respondents claiming that their apparent premonition had come to them during a dream.

One of the most striking experiences was submitted by the parents of a ten-year-old child who perished in the tragedy. The day before the landslide their daughter described dreaming about trying to go to school, but said that there was 'no school there' because 'something black had come down all over it'. In another example, Mrs M.H., a 54-year-old woman from Barnstaple, said that the night before the tragedy she had dreamed that a group of children were trapped in a rectangular room. In her dream, the end of the room was blocked by several wooden bars and the children were trying to climb over the bars. Mrs M.H. was sufficiently worried by the dream to telephone her son and daughter-in-law, and tell them to take special care of their two small daughters. Another respondent, Mrs G.E. from Sidcup, said that a week before the landslide she had dreamed about a group of screaming children being covered by an avalanche of coal, and two months before the tragedy Mrs S.B. from London had dreamed about a school on a hillside, an avalanche and children losing their lives. And so the list went on.

Barker was impressed with his findings and in 1966 set up

the British Premonitions Bureau. The public were asked to submit their alleged premonitions to the Bureau in the hope that Barker would be able to predict, and possibly avert, future tragedies. Unfortunately, his idea didn't catch on. Although his Bureau received about a thousand predictions, the bulk of them came from just six people.[2] Perhaps the strangest story to emerge from the project came from one of these alleged 'precogs', a 44-year-old night telephone operator named Alan Hencher. Hencher usually specialized in predicting air crashes and other major accidents; however, in 1967 he contacted the Bureau to register a far more personal premonition. In what must have been one of the more difficult conversations in the history of parapsychology, Hencher informed Bureau Chief John Barker that Barker would soon die. His comments proved uncannily accurate, with Barker suddenly passing away the following year, aged just 44. To add irony to injury, Barker had previously written a book entitled *Scared To Death*, in which he argued that hearing a premonition of your own demise may induce a deep-seated fear that affects the body's immune system and could result in death. The British Premonitions Bureau closed a few years later due to lack of funds. Apparently, neither Hencher nor any of the other expert precogs foresaw the closure.

Believing that you have seen the future in a dream is surprisingly common, with recent surveys suggesting that around a third of the population experience this phenomenon at some point in their lives. Beliefs like these have been recorded throughout history. The Bible famously describes how Pharaoh dreamed of seven lean cows coming out of a river and eating seven fat cows, and how Joseph interpreted this as the coming of seven years of abundance followed by seven years of famine. The ancient Roman statesman and philosopher Cicero reported having a dream in which he saw 'a noble-looking youth, let down on a chain of gold from the skies'. When he entered the Capitol the following day he saw Octavius and recognized him as the noble-looking youth from his dreams. Octavius later went on to succeed Caesar as Emperor of Rome. In more recent times, Abraham Lincoln reportedly dreamed about an assassination two weeks before being shot dead, Mark Twain described a dream in which he saw his brother's corpse lying in a coffin just a few weeks before his brother was killed in an explosion, and Charles Dickens dreamed of a woman dressed in red called Miss Napier shortly before being visited by a girl wearing a red shawl and introducing herself as Miss Napier.

What could explain these remarkable events? Are people really getting a glimpse of things to come? Can the human psyche really play havoc with the very fabric of time? Is it possible to see tomorrow today?

Throughout history, these questions have taxed the minds of many of the world's greatest thinkers. For example, in about 350 BC the classical Greek philosopher Aristotle penned a short text entitled *On Prophesying by Dreams*. Aristotle's two-part argument was as simple as it was strange. Having thought about the issue for some time, the great philosopher concluded that only God would be able to send prophetic dreams. However, Aristotle had observed that those reporting the dreams did not appear to be especially upstanding citizens, and often turned out to be rather 'commonplace persons'. Figuring that God wouldn't waste time casting his pearls of wisdom among swine like that, Aristotle concluded that prophetic dreams could be safely dismissed as coincidences.

It is an interesting argument, albeit one that is likely to be disputed by both modern scientists and Mrs M.H. from Barnstaple. However, despite over 2,000 years of interest in the mystery of prophetic dreaming, it is only in the last century or so that researchers have managed to solve the puzzle.

Before reading further you might like to make yourself a hot mug of cocoa and snuggle under the covers. We are about to enter the strange world of sleep science.

However, before we begin, let's have a quick memory test. Take a look at the following list of words and try to remember them.

| Lamp | Rock | Apple | Worm | Clock |
| Baby | Horse | Sword | Bird | Desk |

Many thanks, more about this later. Let's start.

Spread Betting

Chapter 5 described how the pioneering work of Eugene Aserinsky helped pave the way for a new science of dreaming. Aserinsky showed that waking up a person after they have spent some time in the REM state is very likely to result in them reporting a dream. In doing so, he kick-started decades of research into the nature of nod. Much of the work involved inviting people to spend the night in special sleep laboratories, monitoring them as they sleep, waking them up after they emerge from REM state, and asking them to describe their dream.[3] The work has yielded many important insights into dreaming. Almost everyone dreams in colour. Those who are blind from birth do not 'see' in their dreams, but experience many more smells, tastes and sounds. Although some dreams are bizarre, many involve everyday chores such as doing the washing-up, filling in tax forms, or vacuuming. If you creep up on someone who is dreaming and quietly play some music, shine a light onto their face or spray them with water, they are very likely to incorporate the stimuli into their dreams. However, perhaps the most important revelation was that you have many more dreams than you might think.

Sleep scientists quickly discovered that you have an average of about four dreams each night. They take place every 90 minutes or so, and each one lasts around 20 minutes. You then forget the vast majority of these episodes when you wake up, leaving you with the impression that you dream far less than is actually the case. The only exception to this rule occurs when you happen to wake up during a dream, perhaps because your alarm clock goes off in the morning or you are disturbed during the night. When this happens you will usually remember the general gist of the dream and perhaps some specific fragments but, unless it is especially striking, you will soon forget all about it. There is, however, a rather unusual set of circumstances that can greatly increase your likelihood of remembering these dreams.

Earlier in this chapter I presented you with a list of ten words and asked you to try to commit them to memory. Now I would like you to attempt to remember all ten words. To help you, here are five words that are associated with a few of the words in the original list.

Light   Time   Fruit   Gallop   Wings

Please get a pen and a piece of paper and try to remember the original list. Don't turn over the page until you have done your best to remember all of the words.

All done? Check your list against page 276.

How did you do? My prediction is that you will have been especially likely to remember the words 'lamp', 'clock', 'apple', 'horse' and 'bird'. Why? Because the associated words 'light', 'fruit', 'time, 'gallop' and 'wings' will have acted as cues. It wasn't that you had forgotten these words, but rather that they were lurking in your unconscious and just required a little help to emerge. A similar principle applies to your memory for dreams. In the same way that the associated words helped you remember words you couldn't instantly recall from the original list, so an event that happens to you when you are awake can trigger the memory of a dream. To discover the relationship between this effect and the gift of prophecy, let's imagine three nights of disturbed dreaming.

On day one you go to bed after a hard day at work. You shut your eyes and slowly lose consciousness. Throughout the night you drift through the various stages of sleep and experience several dreams. At ten past seven your brain once again bursts into action and presents you with another entirely fictitious episode. For the next 20 minutes you find yourself visiting an ice cream factory, falling into a huge vat of raspberry ripple, and attempting to eat your way out. Just when you can take no more, your alarm clock sounds and you wake up with fragments of the factory and raspberry ripple ice cream drifting through your mind.

On day two the same series of events unfolds. You go to bed, drift to sleep and have several dreams. At two o'clock in the morning you are right in the middle of a rather sinister dream in which you are driving along a dark country lane. Eric Chuggers, your all-time favourite rock star is sitting in the passenger seat, and the two of you are chatting easily. Suddenly a giant purple frog jumps out in front of the car, you swerve to avoid the frog but go off the road and hit a tree. However, tonight your cat feels a tad peckish and decides to come and pester you for food. As she jumps onto the bed you wake up from the dream with a vague memory of Eric Chuggers, a giant purple frog, a tree and impending death.

On the third night you again fall asleep. At four o'clock in the morning you experience a rather traumatic dream. It is a surreal affair, with you being forced to audition for the part of an Oompa-Loompa in a new film version of *Charlie and the Chocolate Factory*. Although successful, you subsequently discover that the orange makeup and green hair dye used in the audition is permanent. You suddenly wake up feeling very stressed, remember the audition and spend the next 20 minutes trying to figure out the symbolic meaning of the dream. You then go back to sleep for the rest of the night.

In the morning you wake up, turn on the radio and are shocked to discover that Eric Chuggers was killed in a car accident during the night.

According to the news report, Chuggers was driving through the city, swerved to avoid another car that had drifted onto the wrong side of the road, and collided with a lamppost. Bingo. In the same way that the words 'time' and 'gallop' helped you remember the words 'clock' and 'horse', so the news report acts as a trigger, and the dream about the car accident jumps into your mind. You forget about consuming copious amounts of raspberry ripple ice cream, and the stressful Oompa-Loompa audition. Instead, you remember the one dream that appears to match events in the real world and so become convinced that you may well possess the power of prophecy.

And it doesn't stop there. Soon after convincing yourself that you had a glimpse of the future while fast asleep, a 'let's make this experience as spooky as possible' part of your mind gets to work. Because dreams tend to be somewhat surreal they have the potential to be twisted to match the events that actually transpired. In reality, Eric Chuggers was not driving along a country lane, did not hit a tree and the accident didn't involve a giant purple frog. However, a country lane is similar to a city road, and a lamppost looks a bit like a tree. And what about the giant purple frog? Well, maybe that symbolized something unexpected, such as the car that drifted onto the wrong side of the road. Or maybe it turns out that Chuggers was on hallucinogenic drugs and so might have thought that the oncoming car was indeed a giant purple frog. Or maybe you see a photograph from the scene of the accident and discover that Chuggers' car had a purple mascot on the dashboard. Or maybe an advertising billboard close to the accident contains an image of a giant frog. Or maybe Chuggers' next album was going to have a frog on the cover. Or maybe Chuggers was wearing a purple shirt at the time of the collision. You get the point. Provided that you are creative and want to believe that you have a psychic link with the recently deceased Mr Chuggers, the possibilities for matches are limited only by your imagination.

So far we have focused on your dream about Chuggers because it resembled events that happened a few days later. But let's imagine that instead of Chuggers dying, you went out to a supermarket and were offered an especially gorgeous sample of raspberry ripple ice cream? Under those circumstances you might well have forgotten about the dreams involving Chuggers and the Oompa-Loompas, and been tempted to tell your friends and family about how your dream seemed to have predicted the unexpected encounter with raspberry ripple ice cream. Or let's imagine that a few days later the company that you work for promotes you, and your new position involves wearing a garish uniform. Suddenly the deep symbolism involved in the dream about the Oompa-Loompas would seem obvious, and the dreams about Chuggers and the raspberry ripple ice cream would remain buried in your unconsciousness.

In short, you have lots of dreams and encounter lots of events. Most of the time the dreams are unrelated to the events, and so you forget about them. However, once in a while one of the dreams will correspond to one of the events. Once this happens, it is suddenly easy to remember the dream and convince yourself that it has magically predicted the future. In reality, it is just the laws of probability at work.

This theory also helps explain a rather curious feature of precognitive dreaming. Most premonitions involve a great deal of doom and gloom, with people regularly foreseeing the assassination of world leaders, attending the funeral of close friends, seeing planes falling out of the sky, and watching as countries go to war. People rarely report getting a glimpse of the future and seeing someone deliriously happy on their wedding day or being given a promotion at work. Sleep scientists have discovered that around 80 per cent of dreams are far from sweet, and instead focus on negative events. Because of this, bad news is far more likely than good news to trigger the memory of a dream, explaining why so many precognitive dreams involve foreseeing death and disaster.

At the start of this chapter I described how psychiatrist John Barker found 60 people who appeared to have predicted the Aberfan mining disaster. Does the research into dreaming and memory alter the evidential value of these alleged premonitions? In 36 of Barker's cases the respondents provided no evidence that they had recorded their dream prior to the disaster. These respondents may have had many other dreams before hearing about Aberfan, and then only remembered and reported the one dream that matched the tragedy. Not only that, but the lack of any record made at the time of the dream means that they could have inadvertently twisted and turned the dream to better fit the unfortunate events that transpired. Blackness may have become coal, rooms may have become classrooms, and rolling hillsides may have become a Welsh valley.

Of course, those who believe in paranormal matters might argue that they are convinced by instances when people tell their friends and family about a dream, or describe it in a diary, and then discover that it matches future events. Do these instances constitute a miracle of the mind? To find out, we are going to drift even deeper into the science of sleep.

Interview with Caroline Watt, from the Koester Parapsychology Unit, about sleep precognition
www.richardwiseman.com/paranormality/CarolineWatt.html

'Other Than That, Did You Enjoy the Play, Mrs Lincoln?'

Open almost any book on the paranormal and you will soon discover that President Abraham Lincoln once had one of the most famous precognitive dreams in history. According to the story, in early April 1865 Lincoln went to close friend and bodyguard Ward Hill Lamon and explained that he had recently had a rather unsettling dream. During the dream Lincoln had felt a 'death-like stillness' in his body and heard weeping from a downstairs room in the White House. After searching the building, he arrived at the East Room and came across a corpse wrapped in funeral vestments. A crowd of people were gazing mournfully at the body. When Lincoln asked who had died, he was told that it was the President, and that he had been assassinated.

Two weeks after the dream, Lincoln and his wife went to see a play at Ford's Theatre in Washington D.C. A short time after the start of the play Lincoln was shot dead by Confederate spy John Wilkes Booth.

But the vast majority of books describing the dream aren't giving their readers the full picture. Joe Nickell has had a long and colourful career that has seen him working as an undercover detective, riverboat manager, carnival promoter, and magician. He is now a Senior Research Fellow at the Centre for Inquiry, an American organization that investigates paranormal matters. In the 1990s Nickell decided to take a closer look at Lincoln's apparent prophecy.[4] He tracked down Ward Hill Lamon's account of the incident in his 1895 memoir, *Recollections of Abraham Lincoln*, and discovered that many of the second-hand accounts of the incident missed out one very important part of the episode. After being told about the dream, Lamon expressed his concern, but the President calmly replied, 'In this dream it was not me, but some other fellow, that was killed. It seems that this ghostly assassin tried his hand on some one else.' In other words, Lincoln did not actually think that he had seen his own death but rather that of another President.

Of course, believers might argue that the President did foresee his own assassination, albeit without realizing it. Even assuming that, would the incident count as compelling evidence for precognition? The answer once more lies in the pioneering work of sleep science.

In the late 1960s dream researchers carried out a ground-breaking experiment with a group of patients who were attending therapy sessions to help them cope with the psychological effects of undergoing major surgery.[5] The researchers monitored the patients' dreams over the course of several nights and discovered that when they had attended a therapy session during the day they were far more likely to dream about their medical problems. For example, one patient was having a tough time coping with the drainage tubes resulting from his surgery. After spending time at a therapy session talking about the issue, he was especially likely to have dreams that involved him continually inserting tubes into himself and others. In short, the patients' dreams tended to reflect their anxieties. Similar studies have revealed the same effect. The content of our dreams is not only affected by events in our surroundings, but also often reflects whatever is worrying our minds.

Nickell noted that even the briefest of glances through the history books reveals that Lincoln would have had every reason to be anxious about the possibility of being assassinated. Just before his first inauguration, he was advised to avoid travelling through Baltimore because his aides had uncovered an assassination plot there, and during his time in office he had received several death threats: on one especially memorable occasion an incompetent would-be assassin fired a shot through his top hat. Seen in the light of these findings, Lincoln's famous dream suddenly looks less paranormal.

The same concept may also explain one of the most striking examples of alleged precognition about the Aberfan disaster. At the start of this chapter I described how one of the young girls who would later perish in the tragedy told her parents that she had dreamed about 'something black' coming down over her school and the school no longer being there. For several years before the disaster the local authorities had expressed considerable concern about the wisdom of placing large amounts of mining debris on the hillside, but their worries had been ignored by those running the mine. Correspondence from the time makes the extent of these concerns clear.[6] For example, three years before the disaster, the Borough Engineer in the area wrote to the authorities noting, 'I regard [the situation] as extremely serious as the slurry is so fluid and the gradient so steep that it could not possibly stay in position in the winter time or during periods of heavy rain', and later added, 'this apprehension is also in the minds of . . . the residents in this area as they have previously experienced, during periods of heavy rain, the movement of the slurry to the danger and detriment of people and property'. There is no way of knowing for sure, but it is possible that the young girl's dream may have been reflecting these anxieties.

But what about the other 23 cases in which people produced evidence that they had described their dream before the tragedy occurred, and where the dream did not seem to reflect their anxieties and concerns. To investigate, we need to move away from the science of sleep and into the heady world of statistics. Let's take a closer look at the numbers associated with these seemingly supernatural experiences.

First, let's select a random person from Britain and call him Brian. Next, let's make a few assumptions about Brian. Let's assume that Brian dreams each night of his life from age 15 to 75. There are 365 days in each year, so those 60 years of dreaming will ensure that Brian experiences 21,900 nights of dreams. Let's also assume that an event like the Aberfan disaster will only happen once in each generation, and randomly assign it to any one day. Now, let's assume that Brian will only remember dreaming about the type of terrible events associated with such tragedy once in his entire life. The chances of Brian having his 'disaster' dream the night before the actual tragedy is about a massive 22,000 to 1. Little wonder that Brian would be surprised if it happened to him.

However, here comes the sneaky bit. When Brian is thinking about the chances of the event happening to him he is being very self-centred. In the 1960s there were around 45 million people in Britain, and this same set of events could have happened to any of them. Given that we have already calculated that the chances of any one of them having the 'disaster' dream one night and the tragedy happening the following day is about 22,000 to 1, we would expect 1 person in every 22,000, or roughly 2,000 people, to have this amazing experience in each generation. To say that this group's dreams are accurate is like shooting an arrow into a field, drawing a target around it after it has landed and saying, 'wow, what are the chances of that!'

The principle is known as the 'Law of Large Numbers', and states that unusual events are likely to happen when there are lots of opportunities for that event. It is exactly the same with any national lottery. The chances of any one person hitting the jackpot is millions to one, but still it happens as regular as clockwork each week because such a large number of people buy tickets.

For genuine evidence of premonitions then, the situation is even worse than we have imagined. Our example only concerned people dreaming about the Aberfan tragedy. In reality, national and international bad fortune happens on an almost daily basis. Aeroplane crashes, tsunamis, assassinations, serial killers, earthquakes, kidnappings, acts of terrorism, and so on. Given that people dream about doom and gloom more often than not, the numbers quickly stack up and acts of apparent prophecy are inevitable.

HOW TO CONTROL YOUR DREAMS: PART ONE

Harvard psychologist Daniel Wegner has come up with a simple but effective way of controlling your dreams.[7] As noted in Chapter 4, Wegner has carried out a great deal of work into the so-called 'rebound effect', wherein people who are asked not to think about a certain issue have a surprisingly hard time keeping it out of their mind. Wegner wondered whether the same effect could also be used to influence people's dreams. To find out, he gathered together a group of participants, gave each of them two envelopes, and asked them to open one envelope just before they went to sleep at night and the other when they woke up in the morning.

The first envelope contained an unusual set of instructions. All of the participants were first asked to think of someone that they found especially attractive. Half of the participants were then instructed to spend five minutes trying *not* to think about this person, while the others were asked to think about their dream date. When everyone woke up in the morning they opened the second envelope and found another set of instructions. This time they were asked to describe any dreams that they had had during the night. Wegner discovered that participants who had tried not to think about the person they found attractive were roughly twice as likely as the others to have dreamed about that person. The message is clear – if you want to have a particular person crop up in your dreams, spend five minutes trying *not* to think about that person before you nod off.

So far we have seen how the science of sleep and the study of statistics suggest that precognitive dreams are due to selective remembering, anxiety, and the law of large numbers. Of course, it could always be argued that although these explanations are true of many apparently precognitive dreams, some others are still genuinely supernatural.

The bad news is that although testing this sounds simple in theory, it is tricky in practice. It is no good asking to people to get in touch *after* a national disaster or tragedy because they are likely to report just one of many dreams that they have had, or be part of the group of people who happened to get lucky via the law of large numbers. Also you can't ask people to dream about an event that is in any way predictable. Instead, you have to record lots of people's prophecies *before* an *unpredictable* event has happened. According to the law of large numbers you would end up with a large range of predictions, with just a small sliver of them subsequently proving correct. In contrast, those with a paranormal bent would predict that this would produce a surprisingly large number of the premonitions that point to one particular future.

The good news is that such a study has already been carried out.[8] Welcome to the curious case of Charles Lindbergh, Jr.

'The Biggest Story Since the Resurrection'

Born in 1893, Harvard psychologist Henry Murray spent much of his life trying to unravel the mysteries of the human personality. During the late 1930s he helped develop a well-known psychological tool known as the 'Thematic Apperception Test', or 'TAT' for short. During the TAT people are shown images depicting various ambiguous scenes, such as a mysterious woman looking over a man's shoulder, and asked to describe what they think is happening in the picture ('What do you make of TAT?'). According to proponents of the test, highly trained therapists can use these comments to gain an important insight into people's innermost thoughts, with, for example, remarks about killing, violence and murder all raising red flags. The TAT is not Murray's only claim to fame. Towards the end of the Second World War the American government called on him to help compile a psychological profile of Adolf Hitler. As a face-to-face consultation seemed highly unlikely, Murray was forced to rely on other sources, such as Hitler's school record, writings, and speeches. He concluded that although the dictator appeared outgoing, he was actually quite shy and had a deep-seated need to annex the Sudetenland. Just kidding. Actually, Murray thought that Hitler was a classic example of a 'counteractive narcissist', a man who held grudges, exhibited excessive demands for attention, displayed a tendency to belittle others, and couldn't take a joke. In addition to developing the TAT and putting Hitler on the couch, Murray also conducted a unique test examining the precognitive power of dreams.

In 1927, 25-year-old American Air Mail pilot Charles Lindbergh achieved international fame by making the first solo, non-stop flight across the Atlantic. Two years later Lindbergh married author Anne Spencer Morrow, and the two of them continued to attract huge amounts of publicity by setting several additional flying records, including being the first people to fly from Africa to South America, and pioneering exploration of polar air routes from North America to Asia. In 1930 the Lindberghs had their first child, Charles Lindbergh, Jr., and moved into a large secluded mansion in Hopewell, New Jersey.

On 1 March 1932 the Lindberghs' world changed for ever. At around 10 o'clock at night, the Lindberghs' nurse rushed to Charles Sr. and told him that Charles Jr. had been taken from his room, and that the kidnappers had left a ransom note demanding $50,000. Lindbergh quickly grabbed a gun and searched the grounds. He discovered the homemade ladder that had been used to climb into the child's second storey room, but found no sign of his son. The police were called and Colonel Norman Schwarzkopf (father of General H. Norman Schwarzkopf, who commanded the coalition forces during Operation Desert Shield) took charge of the case and organized a massive search effort. The Lindberghs' fame resulted in the case generating an enormous amount of publicity, with one journalist referring to it as 'the biggest story since the Resurrection'.

A few days after the news of the kidnapping broke, Murray decided to use the high profile case to study the accuracy of precognitive dreaming. He persuaded a national newspaper to ask their readers to submit any premonitions about the case that had appeared in their dreams. Word of Murray's study spread from one newspaper to another, resulting in the psychologist eventually receiving over 1,300 responses. To properly assess the replies, Murray was forced to wait two years until the crime was solved.

Within a few days of his son's disappearance, Lindbergh made various public appeals for the kidnapper to start negotiations. None of them elicited a response. However, when retired schoolteacher John Condon placed an article in a newspaper making it clear that he was willing to act as a go-between and add an additional $1,000 to the ransom, he received a series of notes from the alleged kidnapper. On the second of April, one note asked Condon to meet in a Bronx cemetery and to hand over $50,000 in gold certificates in exchange for information about the child's location. Condon collected the certificates from Lindbergh, handed them over at the meeting, and was told that the child could be found on a boat that was moored along the Massachusetts coast. Lindbergh flew over the region for days but failed to find the alleged boat.

On 12 May 1932, a truck driver pulled over to the side of a road a few miles from the Lindbergh home and walked into a grove of trees to relieve himself. There he chanced upon the corpse of Charles Lindbergh, Jr., buried in a hastily prepared shallow grave. The baby's skull was badly fractured, and his left leg and both hands were missing. A coroner's examination later showed that the baby had been dead for about two months, and that his death was due to a blow on the head.

For over two years the police struggled to solve the crime. Then, in September 1934, a petrol station attendant became suspicious when a customer paid for five gallons of petrol with a ten dollar gold certificate. The attendant took a note of the customer's number plate and passed it on to the authorities. The police identified the vehicle's owner as Bruno Richard Hauptmann, an illegal German immigrant currently working as a carpenter. Police searched Hauptmann's house, discovered $14,000 of the ransom money, and promptly arrested him. During Hauptmann's trial the prosecution showed that his handwriting matched the ransom notes sent to Condon, and that the floorboards in his house were made of the same wood as the ladder discovered at the Lindbergh's house. After an 11-hour deliberation, the jury returned a guilty verdict and Hauptmann was sentenced to death.

Case closed, Murray set to work. He examined his collection of alleged premonitions for three important pieces of information that would have helped the police investigation enormously – the fact that the baby was dead, buried in a grave, and that the grave was near some trees. Only about 5 per cent of the responses suggested that the baby was dead, and only 4 of the 1,300 responses mentioned that he was buried in a grave near some trees. In addition, none of them mentioned the ladder, extortion notes or ransom money. Exactly as predicted by the 'dream premonitions are the work of normal, not paranormal forces' brigade, the respondents' premonitions were all over the place, with only a handful of them containing information that subsequently proved to be accurate. Murray was forced to conclude that his findings did 'not support the contention that distant events and dreams are causally related'. Although people may dream about the future, those dreams do not represent a magical insight into what will be.

Unfortunately, no one seems to have told the public this. In 2009, psychologists Cary Morewedge from Carnegie Mellon University and Michael Norton from Harvard University carried out an experiment to discover whether the modern mind is still attracted to the notion that dreams predict the future.[9] Nearly 200 commuters at a Boston railway station were asked to imagine that they had booked to be on a certain flight, but that the day before they were due to travel one of four events occurred. Either the government issued a warning of a possible terrorist attack, they thought about their plane crashing, a real plane crashed on the same route or that they dreamed about being in an airplane accident. After imagining each scenario, everyone was asked to rate the likelihood of them cancelling their flight. Amazingly, having an alleged precognitive dream came top of the pile, causing a greater sense of anxiety than a government terrorist warning or even an actual crash.

In addition to casting serious doubts on the 'dreams as prophecy' model of the human psyche, the science of sleep has also made considerable progress in tackling perhaps the greatest dream-based mystery of all – what are our dreams actually for?

HOW TO CONTROL YOUR DREAMS: PART TWO

The ultimate type of dream control involves lucid dreaming. This most desirable of night time activities means that you can experience the impossible, allowing you to fly, walk through walls and spend quality time with your favourite celebrity. At first, this strange phenomenon caused a great deal of debate among scientists, with some researchers arguing that perhaps those reporting these experiences weren't actually dreaming. However, the issue was resolved in the late 1970s when dream researcher Keith Hearne monitored the brain activity of those claiming to regularly experience lucid dreams.[10] In perhaps his best-known study, Hearne invited his star subject to his sleep laboratory, asked him to indicate when he was having a lucid dream by moving his eyes right and left eight times, and then monitored his brain activity as he slept. Hearne discovered that the lucid dreams took place during REM sleep and were associated with the same type of brain activity as a normal dream. In short, evidence that lucid dreams are produced by the dreaming brain.

Hearne's work helped kick-start research into lucid dreaming, with scientists investigating a range of issues, including the best ways of increasing the chances of having a lucid dream. Their research suggests that the following steps will help you gain control of your dreams.[11]

1. Set your alarm clock to wake you up about four, six and seven hours after falling asleep. In theory, this will

increase the likelihood of you being woken up during or straight after a dream.

2. If the alarm clock wakes you during a dream, spend ten minutes reading, writing down information about the dream or walking around. Then go back to bed and think about the dream that you had before waking up. Tell yourself that you are going to have the same dream again, but this time you will be aware that you are dreaming.

3. Draw a large letter 'A' (for 'awake') on one of your palms and the letter 'D' (for 'dreaming') on the other. Whenever you notice either of the letters, ask yourself whether you are awake or asleep. This helps you get used to the ritual and therefore asking the same question when you dream. Also, as you prepare to nod off each night, lie in your bed and take a minute to look at the palms of your hands and quietly tell yourself that while you dream you will look at your hands.

4. If you do manage to have a lucid dream, you will find yourself having to decide whether you are dreaming or actually in the real world. The good news is that there are various actions that will allow you to tell fiction from reality. First, try looking into a mirror – in a lucid dream your image will appear blurry. Second, feel free to bite your arm. If you are in a lucid dream you won't be able to feel a thing, whereas in the real world it will hurt like hell. Finally, try leaning against a wall. In a lucid dream you will often fall through the wall, whereas in the real world this will only happen if the building has been constructed by British engineers in the last ten years.

A Stroll Down the Royal Road
to the Unconscious

There is an old joke about a woman who wakes up in the morning, turns to her husband, and says, 'Last night I dreamed that you gave me a wonderful silver necklace for my birthday. What do you think that means?' Her husband replies, 'You'll know tonight.' That evening, the husband returns home with a small package and gives it to his wife. Delighted, she opens the package and finds a copy of *The Interpretation of Dreams* by Sigmund Freud.

The joke is fictitious, but the book is real. Sigmund Freud was fascinated by dreams and famously referred to them as 'the royal road to the unconscious'. Freud's basic model of the mind revolved around the notion that we all have various fears and worries, and that our conscious mind deals with them by repressing them into our unconscious. During dreaming the conscious mind takes a well-earned break, allowing our true desires and emotions to emerge. Freud therefore thought that it was possible to gain an insight into someone's secret desires by having them describe the 'manifest content' (what they actually dreamed about) of a dream, and using this to determine the 'latent content' (the unfulfilled emotions that the dream represents). However, this was often far from straightforward because the unconscious mind isn't big on language and instead tends towards symbolic communication. Although some of these symbols are both universal and obvious (dream 'cigar', think 'penis'), others are very personal and can only be fully understood with the help of a skilled therapist (dream 'constantly hugging a policeman', think '£200 an hour'). Freud's ideas have spawned an entire industry devoted to dream interpretation, with unrepressed salespeople across the world eager to sell manuals, training seminars, and DVDs on the subject. There is just one small problem. Many scientists now think that Freud got it badly wrong, and that these attempts at interpretation are a complete waste of time.

Some scientists adopt a more evolutionary approach to dreaming. If I were to wake you up from REM and ask for a dream report, two things are likely to happen. First, you would probably ask what I am doing in your bedroom. Second, as noted earlier in the chapter, around eight out of ten times you would relate some sort of negative emotion or situation. Perhaps you would say that you were naked in public, sinking in quicksand, or being laughed at by others (or, on a really bad night, all three). Why should such doom and gloom dominate our dreaming mind? According to some evolutionary psychologists, dreams are a dress rehearsal for the threatening situations that you may encounter in the real world.[12] They allow us to think about what to do in difficult situations without actually putting ourselves at risk.

If you are not a fan of this 'dreams are a psychological self-defence class' model, you might be more taken with the ideas proposed by the man who helped unravel the structure of DNA, Francis Crick.[13] In the mid-1980s Crick took a very different approach to the problem, arguing that dreams are the brain's way of sorting through the day's information by throwing out unimportant data and making new connections between events and ideas. Seen from Crick's perspective dreams are both a way of defragmenting the hard drive of the mind, and a giant 'eureka moment' generator. The idea is not without merit, with many great minds reporting that their dreams were a vital source of inspiration. For example, in the 1840s Elias Howe wanted to create the first sewing machine, but couldn't figure out exactly how it would work. One night he dreamed he was surrounded by a group of tribal warriors, and noticed that their spears had holes near their tips. Howe realized that the dream contained the solution to his problem, because by placing a hole at the tip of the needle the thread would catch after it went through cloth and thus make his machine workable. Similarly, chemist August von Kekule spent years trying to figure out the structure of the chemical compound benzene, before dreaming about a snake biting its own tail and realizing that the elusive compound could be composed of a ring of carbon atoms. (Writer Arthur Koestler later described this incident as 'probably the most important dream in history since Joseph's seven fat and seven lean cows'.) The same process has also influenced the history of sport and music, with golfer Jack Nicklaus reporting that he enjoyed a significant upswing in his game after dreaming about a new way to hold his golf club and Paul McCartney noting that the song 'Yesterday' came to him fully formed in a dream. (One academic recently studied McCartney's eureka moment and concluded 'These three components, person, domain and field, comprise a system with circular causality where the individual, the social organization they create within, and the symbol system they use are all equally

important and interdependent in producing creative products. "Yesterday" is but one creative product of this system
at work.'[14] Good to get that sorted.)

If you don't like the idea of dreams as 'threat rehearsal'
or 'idea generator', you might be drawn to the current front-
runner in scientific circles, namely the notion that dreams
are the meaningless product of random brain activity. This
idea, known as the 'activation-synthesis hypothesis', was
first proposed by Harvard psychiatrist James Hobson in the
late 1970s.[15] When you sleep, you obviously don't receive
very much information from your senses. However, according
to Hobson, the evolutionarily older parts of the brain, responsible for basic functions such as breathing and the heartbeat,
produce regular surges in activity that result in random action
throughout the brain. Confused, the more modern part of the
brain does its best to make a meaningful story out of these
sensations, producing bizarre dreams that combine everyday
concerns with random elements. Given that sleep is essential
for your well-being, some theorists believe that in a way,
dreams represent the 'guardians of sleep' – a mechanism that
allows you to deal with brain activity without waking up.
Interestingly, the latest cutting edge research suggests that
they might be right, with people who have damaged the part
of the brain that enables them to dream often reporting
that they find it very difficult to get a good night's sleep.[16] The
'activation-synthesis hypothesis' does not rule out Freud's
notion that dreams reflect everyday worries and concerns, but
it certainly calls into question the idea that they possess a
weird kind of symbolism that can only be unravelled with the
help of a skilled therapist.

Or perhaps it is far simpler than all that. As sleep

researcher Jim Horne from Loughborough University once memorably put it, perhaps dreams are nothing more than a kind of 'cinema of the mind' that is there to keep your brain entertained during the otherwise tedious hours of sleep.

For thousands of years people believed that their dreams could provide a fleeting glimpse of the future. It was not until the 1950s that scientists discovered how to investigate the sleeping brain and figured out the truth about these alleged acts of prophecy. You dream far more than you think and only remember those dreams that appear to come true. Many of your dreams revolve around topics that make you feel anxious and so are more likely to be related to future events. Contrary to popular belief, nearly everyone dreams, and so some of the many millions of dreams that take place each night will depict future events by chance alone. Carry out experiments that eliminate these factors and suddenly your sleeping mind cannot figure out what tomorrow will bring. Perhaps more importantly, these scientific expeditions into the land of nod have produced important clues about the real reasons for your nightly flights of fancy, including how your dreams could prepare you for threatening situations, increase your chances of coming up with creative ideas, and help you get a good night's sleep. There are many more mysteries of sleep waiting to be solved, but one thing is certain – for those who wish to believe in the reality of the paranormal, the findings from the science of sleep are a nightmare.

Conclusion

In which we find out why we are all wired for the weird
and contemplate the nature of wonder.

We are nearing the end of our adventure into the wonderful world of supernatural science. In the first part of our journey we discovered how psychic readings reveal the real you, how out-of-body experiences show how your brain is deciding where you actually are right now, how displays of alleged psychokinesis demonstrate why seeing *isn't* believing, and how attempts to talk with the dead illustrate the power of your unconscious mind. In the second half of our expedition we have discovered how ghostly experiences yield important insights into the psychology of suggestion, how mind control experts manipulate your thoughts, and how prophetic dreams can be explained by the science of sleep. Along the way we have also learned how to create a variety of weird experiences. All being well you should now be able to, amongst other things, conduct a Ouija board session, float away from your body, tell complete strangers all about themselves, appear to bend metal with the power of your mind, avoid being brainwashed, and control your dreams.

There is, however, one important issue that has yet to be tackled. Why is it that we have evolved to experience the impossible? Our minds have helped rid the world of terrible diseases, put a man on the moon, and begun to figure out the origins of the universe. Why then, are they capable of being fooled into thinking that the soul can leave the body, that ghosts exist, and that our dreams really do predict the future?

Strangely enough, the two issues are intimately connected. However, before we find out how this is the case, it is time to return to the exercise that you completed at the very start of the book.

As you may remember, I presented you with an inkblot and asked you to decide what it looked like. This type of test was developed by Freudian therapists in an attempt to gain an insight into their patients. According to them, people unwittingly project their innermost thoughts and feelings onto the image, thus allowing a skilled therapist to gain a deep insight into their patient's unconscious. A considerable amount of research has now demonstrated that such tests are both inaccurate and unreliable.[1] However, every cloud has a silver lining and, on the upside, the test has given rise to several good jokes, including my favourite: 'My psychoanalyst is terrible, and I have no idea what he is doing with so many pictures of my mother naked.'

I digress. Although the test does not provide a portal into your unconscious, it does genuinely measure something that is far more important – your ability to see patterns. How did you score? In the same way that some people are short and others are tall, so some people are naturally good at spotting patterns, even in meaningless inkblots. They look at the image and immediately see the face of a poodle, two rabbits eating some grass, or a teddy bear propped up in bed. Others look at the same image for ten minutes but can still see nothing more than a few black splodges.

The ability to find patterns plays a crucial role in your everyday life because you are constantly required to spot genuine instances of cause and effect. For example, you might feel sick every time you eat certain foods and need to figure out what ingredients are making you ill. Or you might want to buy a new car and so scrutinize several reviews to find the common threads that will lead to an informed purchase. Or you might have to have several relationships before you can work out what makes for your perfect partner. This ability to spot genuine patterns has played a vital role in the success and survival of the human species. Most of the time this skill serves us well and allows us to figure out how the world works. However, once in a while it goes into overdrive and causes us to see what isn't there.

Let's imagine that you are out in the wilderness and the wind causes some nearby bushes to rustle. Moreover, you have been told that there are several hungry tigers in the area and know that they create the same type of rustling sound. You are faced with a simple choice – do you decide that the rustling is due to the wind and stay put, or conclude that it might well be a tiger and run away? Clearly, in terms of your long-term survival, you are better to err on the safe side and come down in favour of the tiger hypothesis. After all, as the old saying goes, it is always better to run away from the wind than face a hungry tiger. Or, to put it in more psychological terms, it is better to see a few patterns that are not actually there than miss one that is.

Because of this, your pattern-finding skills have a built-in tendency to find connections between completely unrelated events. In doing so, you can easily convince yourself that you have experienced the impossible. For example, you might find some striking relationships between a palmist's meaningless statements and your past, and conclude that fortune-telling is genuine. Or you may see correspondences between a random dream and the subsequent events in your life, and decide that you have the gift of prophecy. Or you might look at an unremarkable photograph of rocks reflected in a lake and manage to find a 'ghostly' face in the water. Or you might watch a 'psychic' focus their attention on a spoon, see the spoon bend, and conclude that the bending was the result of the psychic's amazing paranormal abilities. Or you might place a lucky charm in your pocket before an important job interview, be offered the job, and conclude that the charm somehow caused your good fortune. The list is endless.

This grand theory of the paranormal predicts that people who are especially good at finding such patterns should be more likely than most to experience seemingly supernatural phenomena. But is that the case? To find out, researchers presented people with variations on the inkblot test and asked them about the supernatural events that they have experienced.[2] Exactly as predicted, the results revealed that those who obtain especially high scores on pattern-finding tests also experience way more weird stuff.

In short, the ability to find patterns is so important to your survival that your brain would rather see a few imaginary patterns than miss genuine instances of cause and effect. Seen in this way, seemingly supernatural experiences are not the result of your brain tripping up so much as the price you pay for being so amazing the rest of the time.

On Wonder

We have almost completed our journey. It has been fun having you along for the ride, and I hope that you have enjoyed the trip. I would like to leave you with one final thought.

Many years ago I worked as a restaurant magician. Moving from table to table, I would perform card tricks and try my best to ensure that everyone had a good time. At the end of my performance customers would frequently ask the same 'joke' question, namely, 'Can you make my bill disappear?' Each person believed that they were the first one to think of the question and, as the consummate professional, I would force a laugh every time. I was not the only magician to endure the comment night after night. In fact, it was a widely recognized international phenomenon. One well-known American performer wrote the question on a small card and placed a series of ticks next to it. Whenever a customer came out with the question the magician would laugh, then remove the card from his wallet and openly add another tick.

Nowadays I no longer wander around restaurants performing card tricks. However, I do frequently give presentations about the paranormal and talk about much of the material in this book. After the talk at least one person always asks the same question. Instead of wanting to know whether I can make the bill disappear, they ask if there are any paranormal phenomena that I haven't been able to scientifically explain. When I reply that I have yet to come across any compelling evidence for the supernatural, the questioner often looks extremely disappointed. Their reaction usually stems from a belief that a world devoid of supernatural phenomena is somehow less wondrous than one that contains the impossible. I believe that this belief is mistaken.

American mathematician and science writer Martin Gardner was one of my academic heroes. He died in 2010, aged 95, and in one of his last interviews spoke about the notion of wonder.[3] Gardner posed a simple thought experiment. Imagine that someone discovered a river of wine or found a way of making an object float high into the air. How much money would you pay to visit the river or see the levitating object? Most people gladly offer large sums of money to witness such seemingly miraculous phenomena. Gardner then pointed out that a river of water is just as wonderful as a river of wine, and that an object being attracted to the earth is no less remarkable than it being attracted to the sky. I believe that he was right. To believe that the findings of supernatural science remove wonder from the world is to fail to see the remarkable events that surround us every day of our lives. And, unlike those who appear to talk with the dead or move objects with the power of their minds, these amazing phenomena are genuine.

Before we set off on our expedition I said that we were going to travel to a world more wonderful than Oz. There was no need to travel very far. You already live there. As Dorothy so memorably put it at the end of that wonderful movie, there's no place like home.

The Instant Superhero Kit

I thought it would be fun to leave you with a parting gift. I have put together a set of quick and quirky psychological demonstrations that you can use to impress your friends, family and colleagues. These demonstrations are based on the theories and ideas encountered on our trip, and are designed to act as an inspiring memento of our journey. They will only take a few moments to learn, and together they form 'The Instant Superhero Kit'. Enjoy.

The Reading

Chapter 1 examined how it is that psychics, mediums and astrologers appear to give highly accurate and impressive readings. It takes practice to master the psychological principles involved in a professional 'cold reading'. However, you can instantly use the following demonstration to convince complete strangers that you know all about them.

In the late 1940s psychologist Bertram Forer carried out a groundbreaking experiment in which he gave each of his students exactly the same personality description and discovered that almost all of them rated it as being highly accurate.[1] This phenomenon, now known as 'The Barnum Effect', can be used to give the impression that you have a deep and mysterious insight into a stranger's personality.

To concoct a convincing cover story, first find out if the person you are trying to impress is into palmistry, astrology or psychology. Then look at their hand, ask for their date or birth, or have them draw a house, and recite the following:

> I get the impression that you are a loyal and devoted friend – someone that people can rely on in times of difficulty. Although you are fair-minded, you are also a far more ambitious person than your friends and colleagues realize. Most of the time you give the impression of being strong, but deep down you sometimes worry about what the future will bring. You are

the sort of person who endorses very general statements about themselves. (Just kidding. Sorry if you
read that out.) I have a feeling that in certain circumstances you can be something of a perfectionist, and
that this sometimes annoys those around you. You are
good at seeing both sides of an argument rather than
jumping to conclusions. You are the type of person
that likes to gather together all of the facts and then
make a decision. Is that right? When you look back
on your life you sometimes dwell on what you might
have done differently, but in general you focus on the
future. Although you enjoy change and variety, you
are also attracted to a sense of routine and stability.
You are facing a significant decision right now, or
have recently experienced a large change in your life.
You know that you have considerable unused capac-
ity that you have not turned to your advantage, and
at times you are extroverted and sociable, while at
other times you are far more introverted and reserved.

Science predicts that the stranger will be mightily impressed.
That is, of course, unless they have also read this book.

Instant Anaesthetist

Chapter 2 delved deep into the science behind out-of-body experiences and discovered that these strange sensations provided a unique insight into how your brain figures out where 'you' are every moment of your waking life. Some of the research in this area has explored how your brain uses visual information to decide where 'you' are by conducting studies in which people feel as if a rubber hand, or even a tabletop, is part of them. This 'anaesthetized finger' demonstration is conceptually identical to these experiments. Ask a friend to extend their right index finger. Now, extend your left index finger and clasp your hands together so that your and your friend's index fingers touch along their length (see the photograph below).

Next, ask your friend to use the thumb and first finger of their left hand to stroke along the sides of this 'double finger'. Have them rub their left thumb along the front of their right index finger and their left index finger along the front of your left index finger. Something very strange will happen. Your friend will feel as if their left index finger has become completely numb.

Your friend's brain sees what it believes to be their left index finger being stroked, but feels nothing, and decides that the finger must be numb. In addition to illustrating the innermost workings of the brain, this demonstration is great for chatting people up in bars.

The Suggestibility Test

Chapter 4 revealed how investigations into table-turning, the Ouija board and automatic writing led to the discovery of a form of unconscious movement known as 'ideomotor action'. Suggestible people are especially prone to the ideomotor action and you can use the following exercise to assess your friend's level of suggestibility.

Ask your friend to hold out their arms in front of them, ensuring that their arms are parallel to the ground and that both of their hands are face down and level. Now ask them to close their eyes while you read out the following paragraph, slowly and clearly:

> I am going to take you through a simple visualization exercise. First of all, imagine a heavy stack of books being tied together with some thick string, and that the end of the string is attached to the fingers of your left hand. The books are hanging under your left hand and tugging down on your arm, pulling it towards the ground. Don't consciously move your hands, but instead just listen to my voice and let the images flow through your mind. Imagine the weight of the books gently pulling your left arm towards the ground, feeling heavier and heavier as time goes on. Now imagine a balloon filled with helium and attached to a thin thread. The end of the thread is tied to the fingers of

your right hand and is gently pulling your hand into the air. The books are dragging your left hand down towards the ground and the balloon is pulling your right hand towards the ceiling. Don't consciously move your hands, but instead just listen to my voice and let the images flow through your mind. Your left hand being pulled down and your right hand being pulled up. Excellent. Now open your eyes and relax your arms.

Look at the position of your friend's hands at the end of the exercise. The hands started at the same level. Has the left hand moved lower, and the right higher? If they are still level, or just a few inches apart, then the person is not especially suggestible. If the person's hands have moved more than a couple of inches apart then they are the more suggestible type. In addition to assessing their level of suggestibility, the test will also reveal an insight into their character. Non-suggestible types tend to be more down-to-earth, logical, and enjoy puzzles and games. In contrast, suggestible types tend to have a good imagination, be sensitive, intuitive, and find it easier to become absorbed in books and films.

Me performing the suggestibility test
www.richardwiseman.com/paranormality/SuggestTest.html

Mind Over Matter

Chapter 3 investigated how those claiming to be able to move objects with the power of their mind reveal that you are only seeing a small fraction of what is actually taking place in front of your eyes. This important psychological principle is illustrated in the following two-part demonstration. All you need is a plastic straw, a plastic bottle and a table.

Seconds before you begin, secretly rub the straw on your clothing to ensure that it builds up a static charge. Next, carefully balance the straw horizontally across the top of a plastic bottle (see photograph).

Announce that you seem to have acquired some very odd paranormal powers, place your right hand about an inch away from one end of the straw, and rub your fingers together. The straw will magically rotate on the bottle top, moving towards your fingers.

For the second part of the performance, place the straw on a tabletop a few inches from the edge of the table. The straw needs to be lying on its side and parallel to your body. Once again, rub the tips of your fingers together as if you are trying to summon your latent powers. Now place your right hand on the tabletop a few inches beyond the straw (see the photograph below).

Next, tilt your head down slightly as you focus your attention on the straw. Slowly rub your fingers together and, at the same time, secretly blow towards the surface of the table. The air currents will travel along the table and move the straw.

Voilà, an instant miracle.

Using two different methods (static electricity and blowing) to obtain the same effect is an important principle in faking mind over matter. Similarly, during the second part of the demonstration, people's attention is directed towards your fingers and away from your mouth, which also helps misdirect them away from the real source of the movement.

THE INSTANT SUPERHERO KIT

Me performing the straw demonstrations
www.richardwiseman.com/paranormality/PKdemo.html

The Ritual

Chapter 5 ventured deep into the spooky world of ghosts and hauntings, and discovered how things that go bump in the night are actually due to the psychology of suggestion, a heightened sense of fear causing hyper-vigilance, and the brain's 'Hypersensitive Agency Detection Device'. Many people would love to experience a ghost, and this demonstration will convince your friends that you have the power to summon the spirits.

Ask your friend to stand about half a metre in front of a large mirror. Next, place a candle or other dim light directly behind them, and then turn off the lights. After about a minute of them gazing at their reflection, they will start to experience a strange illusion. According to work conducted by Italian psychologist Giovanni Caputo,[2] about 70 per cent of people will see their face become horribly distorted, with many eventually seeing it contort into the face of another person. According to folklore, the effect is enhanced if your friend chants the words 'Bloody Mary' 13 times. Although researchers are not sure what produces the weird effect, it seems to be due to the procedure preventing your brain 'binding' together the different features of your face into a single image.

Finish the demonstration by explaining that it is quite likely that the spirits will now follow them home and give them terrible nightmares for a week (especially effective if their hands were far apart during the suggestibility test).

Control Freak

Chapter 6 explored the world of mind control, revealing how remarkable displays of telepathy led to the discovery of muscle reading, and how the study of cult leaders revealed the power of persuasion. Starting a cult is probably not a very good idea. There are, however, a few fun ways in which you can appear to control your friend's behaviour.

First, ask your friend to clasp their hands together but to keep the index fingers of each hand extended, with a gap of about an inch between the two fingertips (see photograph below).

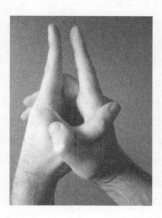

Next, announce that you are going to use the power of your mind to make their fingers drift together. Ask your friend to try as hard as they can to keep their index fingers apart, but to imagine a fine thread being wrapped around the ends, and the loop slowly tightening. You might find it helpful to mime the wrapping and tightening of the thread. After a few seconds your friend's muscles will become fatigued and their fingers will slowly drift together.

Second, ask your friend to place their right hand flat on a tabletop. Their thumb and fingers should be spread out and flat on the table. Ask them to bend the second finger of their right hand inwards at the second joint and lay it against the table (see photograph).

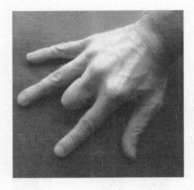

Announce that you will use your mental abilities to prevent them lifting the third finger of their right hand off the table. Try as they might, your friend will not be able to move their third finger.

I hope that you enjoy demonstrating your newfound super-powers and will use them as a force for good.

Notes

Introduction

1 My experiment with Jaytee is described in:
R. Wiseman, M. Smith, J. Milton (1998). 'Can animals detect when their owners are returning home? An experimental test of the "psychic pet" phenomenon.' *British Journal of Psychology,* 89, 453–62.
Rupert Sheldrake has also conducted research with Jaytee and believes that the results provide evidence for psychic ability. This work is described in his book *Dogs That Know When Their Owners are Coming Home.* My response to these studies is available at www.richardwiseman.com/jaytee.

2 L. J. Chapman and J. P. Chapman (1967). 'Genesis of popular but erroneous psychodiagnostic observations'. *Journal of Abnormal Psychology*, 72, pages 193–204.

3 D. A. Redelmeier and A. Tversky (1996). 'On the belief that arthritis pain is related to the weather'. *Proc Natl Acad Sci USA*, 93, pages 2895–6.

1. FORTUNE-TELLING

1 Much of the information in this section is taken from:
M. J. Mooney (2009). 'The Demystifying Adventures of the Amazing Randi'. *SF Weekly News*, August 26.
(http://www.sfweekly.com/2009-08-26/news/the-demystifying-adventures-of-the-amazing-randi/1/)

2 For more information about this test, see:
   http://www.guardian.co.uk/science/2009/may/12/psychic-claims-
   james-randi-paranormal

3 Patricia Putt later complained about the conditions associated
   with the test. Her remarks, and my commentary on them, can be
   seen here:
   http://richardwiseman.wordpress.com/2009/05/27/patricia-putt-
   replies/

4 H. G. Boerenkamp (1988). A *Study of Paranormal Impressions
   of Psychics*. CIP-Gegevens Koninklijke, The Hague.
   This work was also published in a series of articles in the
   European Journal of Parapsychology from 1983 to 1987.

5 S. A. Schouten, (1994). 'An overview of quantitatively evaluated
   studies with mediums and psychics'. *The Journal of the
   American Society for Psychical Research*, 88, pages 221–54.

6 C. A. Roe (1998). 'Belief in the paranormal and attendance at
   psychic readings'. *Journal of the American Society for Psychical
   Research*, 90, pages 25–51.

7 For more information about cold reading, see:
   I. Rowland (1998). *The Full Facts Book of Cold Reading*. Ian
   Rowland Limited, London.

8 For a review of this literature, see:
   D. G. Myers (2008). *Social Psychology*. McGraw-Hill Higher
   Education, New York.

9 A. H. Hastorf and H. Cantril (1954). 'They Saw a Game: A Case
   Study'. *Journal of Abnormal and Social Psychology*, 49, pages
   129–34.

10 D. H. Naftulin, J. E. Ware and F. A. Donnelly (1973). 'The
   Doctor Fox Lecture: A Paradigm of Educational Seduction'.
   Journal of Medical Education, 48, pages 630–5.

11 The Editors of *Lingua Franca* (2000). *The Sokal Hoax: The
   Sham That Shook the Academy*. Bison Books, Lincoln, NE.

12 G. A. Dean, I. W. Kelly, D. H. Saklofske and A. Furnham (1992).

'Graphology and human judgement'. In *The Write Stuff* (ed. B. Beyerstein and D. Beyerstein), pages 349–95. Prometheus Books, Buffalo, NY.

13 A. C. Little and D. I. Perrett (2007). 'Using composite face images to assess accuracy in personality attribution'. *British Journal of Psychology*, 98, pages 111–26.

14 Figures reproduced with permission from *The British Journal of Psychology* © The British Psychological Society.

15 For more information about population stereotypes, see: D. Marks (2000). *The Psychology of the Psychic (2nd Edition)*. Prometheus Books, Amherst, NY.

16 S. J. Blackmore (1997). 'Probability misjudgment and belief in the paranormal: A newspaper survey'. *British Journal of Psychology*, 88, pages 683–9.

17 B. Jones (1989). *King of the Cold Readers: Advanced professional pseudo-psychic techniques*. Jeff Busby Magic Inc., Bakersfield, CA.

18 B. Couttie (1988). *Forbidden Knowledge: The Paranormal Paradox*. Lutterworth Press, Cambridge.

19 W. F. Chaplin, J. B. Phillips, J. D. Brown, N. R. Clanton and J. L. Stein (2000). 'Handshaking, gender, personality and first impressions'. *Journal of Personality and Social Psychology*, 79, pages 110–17.

2. OUT-OF-BODY EXPERIENCES

1 C. A. Alvarado (2000). 'Out-of-body experiences'. In *Varieties of anomalous experiences* (ed. E. Cardeña, S. J. Lynn and S. Krippner), pages 183–218. American Psychological Association, Washington D.C.

2 G. Gabbard and S. Twemlow (1984). *With the eyes of the mind*. Praeger Scientific, New York.

3 For further information about Mumler, see:
   L. Kaplan (2008). *The Strange Case of William Mumler, Spirit Photographer*. University of Minnesota Press, MN.

4 For further information about photographing the soul see:
   H. Carrington and J. R. Meader (1912). *Death, its Causes and Phenomena*. Rider, London.

5 M. Willin (2007). *Ghosts Caught on Film: Photographs of the Paranormal?* David & Charles, Cincinnati.

6 D. MacDougall (1907). 'Hypothesis concerning soul substance, together with experimental evidence of the existence of such substance'. *Journal of the American Society for Psychical Research*, 1, pages 237–44.

7 M. Roach (2003). *Stiff: The Curious Lives of Human Cadavers*. W. W. Norton, New York.

8 An account of the experiments carried out by Watters and Hopper can be found in:
   S. J. Blackmore (1982). *Beyond the Body: An Investigation into Out-of-the-Body Experiences*. Paladin Grafton Books, London.

9 K. Clark (1984). 'Clinical Interventions with Near-Death Experiencers.' In *The Near-Death Experience: Problems, Prospects, Perspectives* (ed. B. Greyson and C. P. Flynn), pages 242–55. Charles C. Thomas, Springfield, IL.

10 E. Hayden, S. Mulligan and B. L. Beyerstein (1996). 'Maria's NDE: Waiting for the Other Shoe to Drop'. *Skeptical Inquirer*, 20(4), pages 27–33.

11 This questionnaire is based upon work described in:
    A. Tellegen and G. Atkinson (1974). 'Openness to absorbing and self-altering experiences ("absorption"), a trait related to hypnotic susceptibility'. *Journal of Abnormal Psychology*, 83, pages 268–77.

12 K. Osis (1974). 'Perspectives for out-of-body research'. In *Research in Parapsychology* (ed. W. G. Roll, R. L. Morris and J. D. Morris, 1973), pages 110–13.

13 J. Palmer and R. Lieberman (1975). 'The influence of
    psychological set on ESP and out-of-body experiences'. *Journal
    of the American Society for Psychical Research*, 69, pages
    235–43.
    J. Palmer and C. Vassar (1974). 'ESP and out-of-body
    experiences: An exploratory study'. *Journal of the American
    Society for Psychical Research*, 68, pages 257–80.

14 M. Botvinick and J. Cohen (1998). 'Rubber hands "feel" touch
    that eyes see'. *Nature*, 391, page 756.

15 G. L. Moseley, et al. (2008). 'Psychologically induced cooling of
    a specific body part caused by the illusory ownership of an
    artificial counterpart'. *Proc Natl Acad Sci*, 105, pages 13169–73.

16 S. Blakeslee and V. S. Ramachandran (1998). *Phantoms in the
    Brain: Human Nature and the Architecture of the Mind*. William
    Morrow, New York.
    K. C. Armel and V. S. Ramachandran (2003). 'Projecting
    sensations to external objects: Evidence from skin conductance
    response'. *Proceedings of the Royal Society of London:
    Biological*, 270, pages 1499–506.

17 V. S. Ramachandran and D. Rogers-Ramachandran (1996).
    'Synaesthesia in phantom limbs induced with mirrors'. *Proc R
    Soc Lond B Biol Sci*, 263, pages 286–377.

18 B. Lenggenhager, T. Tadi, T. Metzinger and O. Blanke (2007).
    'Video ergo sum: Manipulation of bodily self consciousness'.
    Science. 317, pages 1096–9.

19 E. L. Altschuler and V. S. Ramachandran (2007). 'A simple
    method to stand outside oneself'. *Perception*, 36(4), pages 632–4.

20 S. J. Blackmore and F. Chamberlain (1993). 'ESP and Thought
    Concordance in Twins: A Method of Comparison'. *Journal of
    the Society for Psychical Research*. 59, pages 89–96.

21 S. J. Blackmore (1987). 'Where am I?: Perspectives in imagery,
    and the out-of-body experience'. *Journal of Mental Imagery*, 11,
    pages 53–66.

3. MIND OVER MATTER

1 For further information about Hydrick, see:
D. Korem (1988). *Powers: Testing the psychic & supernatural.*
InterVarsity Press, Downers Grove, IL.
'Psychic Confession', a documentary made by Korem about his
time with Hydrick.
J. Randi (1981). '"Top Psychic" Hydrick: Puffery and Puffs'.
The Skeptical Inquirer, 5(4), pages 15–18.

2 D. Korem and P. D. Meier (1981). *The Fakers: Exploding the
myths of the supernatural.* Baker Book House, Grand Rapids,
MN.

3 R. Beene (1989). '"Sir James" molest suspect says he's
misunderstood, but prosecutors insist he's a con man'. *LA
Times*, February 2006.

4 This test is based on a similar task described in:
L. Wardlow Lane, M. Groisman and V. S. Ferreira (2006). 'Don't
talk about pink elephants! Speakers' control over leaking private
information during language production'. *Psychological Science*,
17, pages 273–7.

5 J. Steinmeyer (2006). *Art and Artifice and Other Essays of
Illusion.* Carroll & Graf, New York.

6 B. Singer and V. A. Benassi (1980–81). 'Fooling some of the
people all of the time'. *Skeptical Inquirer*, 5(2), pages 17–24.

7 R. Hodgson and S. J. Davey (1887). 'The possibilities of
malobservation and lapse of memory from a practical point of
view'. *Proceedings of the Society for Psychical Research*, 4, pages
381–404.

8 A. R. Wallace (1891). Correspondence: 'Mr S. J. Davey's
Experiments'. *Journal of the Society for Psychical Research*, 5,
page 43.

9 R. Hodgson (1892). 'Mr. Davey's imitations by conjuring of
   phenomena sometimes attributed to spirit agency'. *Proceedings
   of the Society for Psychical Research*, 8, pages 252–310.
10 Images reproduced by permission of J. Kevin O'Regan,
   Laboratoire Psychologie de la Perception CNRS, Université Paris
   Descartes.
11 R. Wiseman and E. Haraldsson (1995). 'Investigating macro-PK
   in India: Swami Premananda'. *Journal of the Society for
   Psychical Research*, 60, pages 193–202.
12 H. Münsterberg (1908). *On the Witness Stand: Essays on
   Psychology and Crime*. Page & Co., Doubleday, New York.
13 R. Buckhout (1974). 'Eyewitness testimony'. *Scientific American*,
   231, pages 23–31.
14 R. Buckhout (1975). 'Nearly 2000 witnesses can be wrong'.
   Social Action and the Law, 2, page 7.

4. TALKING WITH THE DEAD

1 For further information about the Fox sisters, see:
   B. Weisberg (2004). *Talking to the Dead: Kate and Maggie Fox
   and the Rise of Spiritualism*. HarperSanFrancisco, San Francisco.
2 P. Lamont (2004). 'Spiritualism and a mid-Victorian crisis of
   evidence'. *Historical Journal*, 47(4), pages 897–920.
3 For a comprehensive account of the confession, see:
   R. B. Davenport (1888). *The Death-Blow to Spiritualism: being
   the true story of the Fox sisters, as revealed by authority of
   Margaret Fox Kane and Catherine Fox Jencken*. G. W.
   Dillingham, New York.
4 P. P. Alexander (1871). *Spiritualism: a narrative with a
   discussion*. William Nimmo, Edinburgh.
5 N. S. Godfrey (1853). *Table Turning: the Devil's Modern
   Masterpiece; Being the Result of a Course of Experiments*.
   Thames Ditton, UK.

6 D. Graves (1996). *Scientists of Faith*. Kregel Resources, Grand Rapids, MI.

7 M. Faraday (1853). 'Experimental investigation of table moving'. *Athenaeum*, 1340, pages 801–3.

8 J. Jastrow (1900). *Fact and Fable in Psychology*. Houghton Mifflin Company, New York.

9 For a review of this work, see:
E. Jacobson (1982). *The Human Mind: A physiological clarification*. Charles C. Thomas, Springfield, IL.

10 H. H. Spitz (1997). *Nonconscious Movements: From Mystical Messages to Facilitated Communication*. Lawrence Erlbaum Associates, Princeton, NJ.

11 D. M. Wegner and D. J. Schneider (2003). 'The White Bear Story'. *Psychological Inquiry*, 14, pages 326–29.

12 O. P. John and J. J. Gross (2004). 'Healthy and unhealthy emotion regulation: Personality processes, individual differences, and life span development'. *Journal of Personality*, 72, pages 1301–17.
A. G. Harvey (2003). 'The attempted suppression of presleep cognitive activity in insomnia'. *Cognitive Therapy and Research*, 27, pages 593–602.

13 D. M. Wegner, M. E. Ansfield and D. Pilloff (1998). 'The putt and the pendulum: Ironic effects of the mental control of action'. *Psychological Science*, 9, pages 196–9.

14 F. C. Bakker, R. R. D. Oudejans, O. Binsch and J. Van der Kamp (2006). 'Penalty shooting and gaze behavior: Unwanted effects of the wish not to miss'. *International Journal of Sport Psychology*, 37, pages 265–80.

15 J. Etkin (2001). 'Erratic Pitching – performance anxiety of baseball players'. *Baseball Digest*, August 2001, pages 52–6.

16 W. F. Prince (1964). *The Case of Patience Worth*. University Books, Inc., New York.

17 D. Wegner (2002). *The Illusion of Conscious Will*. The MIT Press, Cambridge, MA.

NOTES

18 B. Libet, C. A. Gleason, E. W. Wright and D. K. Pearl (1983).
    'Time of conscious intention to act in relation to onset of
    cerebral activity (readiness-potential). The unconscious initiation
    of a freely voluntary act'. *Brain*, 106, pages 623–42.
    B. Libet (1985). 'Unconscious cerebral initiative and the role of
    conscious will in voluntary action'. *Behavioral and Brain
    Sciences*, 8, pages 529–66.
19 Described in 'Time and the Observer' by D. C. Dennett and
    M. Kinsbourne, in *The Nature of Consciousness: Philosophical
    debates*, (Ned Block, Owen Flanigan, et al., eds., 1997), The
    MIT Press, Cambridge, MA, page 168.

INTERMISSION

1 For further information about Gef, see:
   H. Price (1936). *Confessions of a Ghost-Hunter*. Putnam & Co.
   Ltd, London.
   H. Price and R. S. Lambert (1936). *The Haunting of Cashen's
   Gap: A Modern 'Miracle' Investigated*. Methuen & Co. Ltd.,
   London.

5. GHOST-HUNTING

1 D. P. Musella (2005). 'Gallup poll shows that Americans' belief
   in the paranormal persists'. *Skeptical Inquirer*, 29(5), page 5.
2 R. Lange, J. Houran, T. M. Harte and R. A. Havens (1996).
   'Contextual mediation of perceptions in hauntings and
   poltergeist-like experiences'. *Perceptual and Motor Skills*, 82,
   pages 755–62.
3 D. J. Hufford (1982). *The Terror That Comes in the Night*.
   University of Pennsylvania Press, Philidelphia.

T. Kotorii, N. Uchimura, Y. Hashizume, S. Shirakawa, T. Satomura et al. (2001). 'Questionnaire relating to sleep paralysis'. *Psychiatry and Clinical Neurosciences*, 55, pages 265–6.

4 C. Brown (2003). 'The stubborn scientist who unraveled a mystery of the night'. *Smithsonian Magazine*, October 2003.

5 E. Aserinsky and N. Kleitman (1953). 'Regularly occurring periods of eye motility, and concomitant phenomena, during sleep'. *Science*, 118, pages 273–4.

6 For additional information about this work, see:
R. Wiseman, C. Watt, E. Greening, P. Stevens and C. O'Keeffe (2002). 'An investigation into the alleged haunting of Hampton Court Palace: Psychological variables and magnetic fields'. *Journal of Parapsychology*, 66(4), pages 387–408.
R. Wiseman, C. Watt, P. Stevens, E. Greening and C. O'Keeffe (2003). 'An investigation into alleged "hauntings"'. *The British Journal of Psychology*, 94, pages 195–211.

7 G. W. Lambert (1955). 'Poltergeists: a physical theory'. *Journal of the Society for Psychical Research*, 38, pages 49–71.

8 A. Gauld and A. D. Cornell (1979). *Poltergeists*. Routledge & Kegan Paul, London.

9 A. Cornell (1959). 'An experiment in apparitional observation and findings'. *Journal of the Society for Psychical Research*, 40, pages 120–4.
A. Cornell (1960). 'Further experiments in apparitional observations'. *Journal of the Society for Psychical Research*, 40, pages 409–18.

10 V. Tandy and T. Lawrence (1998). 'The ghost in the machine'. *Journal of the Society for Psychical Research*, 62, pages 360–4.

11 V. Tandy (2000). 'Something in the cellar'. *Journal of the Society for Psychical Research*, 64, pages 129–40.

12 C. M. Cook and M. A. Persinger (1997). 'Experimental induction of the "sense presence" in normal subjects and an exceptional subject'. *Perceptual and Motor Skills*, 85, pages 683–93.

C. M. Cook and M. A. Persinger (2001). 'Geophysical variables and behavior: XCII. Experimental elicitation of the experience of a sentient being by right hemispheric, weak magnetic fields: Interaction with temporal lobe sensitivity'. *Perceptual and Motor Skills*, 92, pages 447–8.

13 P. Granqvist, M. Fredrikson, P. Unge, A. Hagenfeldt, S. Valind, D. Larhammar and M. Larsson (2005). 'Sensed presence and mystical experiences are predicted by suggestibility, not by the application of transcranial weak complex magnetic fields'. *Neuroscience Letters*, 379, pages 1–6.

M. Larsson, D. Larhammar, M. Fredrikson and P. Granqvist (2005). 'Reply to M.A. Persinger and S. A. Koren's response to Granqvist et al. "Sensed presence and mystical experiences are predicted by suggestibility, not by the application of transcranial weak complex magnetic fields"'. *Neuroscience Letters*, 380, pages 348–50.

For additional information about this work, see: http://www.nature.com/news/2004/041206/full/news041206-10.html

14 C. C. French, U. Haque, R. Bunton-Stasyshyn and R. Davis (2009). 'The "Haunt" Project: An attempt to build a "haunted" room by manipulating complex electromagnetic fields and infrasound'. *Cortex*. 45, pages 619–29.

For further information about the possible relationship between hauntings and electromagnetism, see:

J. J. Braithwaite (2008) 'Putting magnetism in its place: A critical examination of the weak-intensity magnetic field account for anomalous haunt-type experiences'. *Journal for the Society of Psychical Research*, 890, pages 34–50.

J. J. and M. Townsend (2005). 'Sleeping with the entity: A quantitative magnetic investigation of an English castle's reputedly haunted bedroom'. *European Journal of Parapsychology*. 20.1, pages 65–78.

15 E. E. Slosson (1899). 'A lecture experiment in hallucinations'. *Psychology Review*, 6, pages 407–8.

16 M. O'Mahony (1978). 'Smell illusions and suggestion: Reports of smells contingent on tones played on television and radio'. *Chemical Senses and Flavour*, 3, pages 183–9.

17 R. Lange and J. Houran (1999). 'The role of fear in delusions of the paranormal'. *Journal of Nervous and Mental Disease*, 187, pages 159–66.

18 R. Lange and J. Houran (1997). 'Context-induced paranormal experiences: Support for Houran and Lange's model of haunting phenomena'. *Perceptual and Motor Skills*, 84, pages 1455–8.

19 J. Houran and R. Lange (1996). 'Diary of events in a thoroughly unhaunted house'. *Perceptual and Motor Skills*, 83, pages 499–502.

20 Much of the information in this section is based on a report of Smyth's work in the 1970s BBC documentary series, *Leap in the Dark*.

21 J. M. Bering (2006). 'The cognitive psychology of belief in the supernatural'. *American Scientist*, 94, pages 142–9.

22 J. L. Barrett (2004). *Why Would Anyone Believe in God?* AltaMira Press, Lanham, MD.

6. MIND CONTROL

1 For further information about Bishop, see:
H. H. Spitz (1997). *Nonconscious Movements: From mystical messages to facilitated communication*. Lawrence Erlbaum Associates, Princeton, NJ.
R. Jay (1986). *Learned Pigs and Fireproof Women*. Robert Hale, London.
B. H. Wiley (2009). 'The Thought-Reader Craze'. *The Conjuring Arts Research Center*. 4(1), pages 9–134. Gibeciere, NY.

2 For more information about Clever Hans, see:
   H. H. Spitz (1997). *Nonconscious Movements: From Mystical Messages to Facilitated Communication*. Lawrence Erlbaum Associates, Princeton, NJ.
   O. Pfungst (1911). *Clever Hans (The horse of Mr. von Osten): A contribution to experimental animal and human psychology*. Henry Holt, New York.

3 R. Rosenthal and K. Fode (1963). 'The effect of experimenter bias on the performance of the albino rat'. *Behavioral Science*, 8, pages 183–9.

4 R. Rosenthal and L. Jacobson (1968). *Pygmalion in the Classroom: Teacher expectations and pupils' intellectual development*. Holt, Rinehart and Winston, New York.

5 G. L. Wells (1988). *Eyewitness Identification: A system handbook*. Carswell, Toronto.

6 H. B. Gibson (1991). 'Can hypnosis compel people to commit harmful, immoral and criminal acts?: A review of the literature'. *Contemporary Hypnosis*, 8, pages 129–40.

7 M. T. Orne and F. J. Evans (1965). 'Social control in the psychological experiment: Antisocial behavior and hypnosis'. *Journal of Personality and Social Psychology*, 1, pages 189–200.

8 For further information about Jim Jones, see:
   J. Mills (1979). *Six Years with God*. A&W Publishers, New York.
   D. G. Myers (2010). *Social Psychology* (10th ed.). McGraw-Hill, New York.

9 J. L. Freedman and S. C. Fraser (1966). 'Compliance without pressure: The foot-in-the-door technique'. *Journal of Personality and Social Psychology*, 4, pages 196–202.

10 S. E. Asch (1951). 'Effects of group pressure upon the modification and distortion of judgment'. In *Groups, Leadership and Men* (ed. H. Guetzkow). Carnegie Press, Pittsburgh, PA.

11 E. Aronson and J. Mills (1959). 'The effect of severity of

initiation on liking for a group'. *Journal of Abnormal and Social Psychology*, 59, pages 177–81.

12 L. Festinger, H. W. Riecken and S. Schachter (1956). *When Prophecy Fails: A Social and Psychological Study of a Modern Group that Predicted the Destruction of the World*. University of Minnesota Press, Minneapolis, MN.

7. PROPHECY

1 J. C. Barker (1967). 'Premonitions of the Aberfan Disaster'. *Journal of the Society for Psychical Research*, December 1967, 44, pages 168–81.

2 A. MacKenzie (1974). *The Riddle of the Future: A modern study of precognition*. Arthur Barker, London.

3 A. M. Arkin, J. S. Antrobus and J. Ellman (1978). *The Mind in Sleep: Psychology and psychophysiology*. Erlbaum, New Jersey.

4 J. Nickell (1999). 'Paranormal Lincoln'. *Skeptical Inquirer*, 23, 127–31.

5 L. Breger, I. Hunter and R. W. Lane (1971). *The Effect of Stress on Dreams*. International Universities Press, New York.

6 Taken from:
http://www.nuffield.ox.ac.uk/politics/aberfan/dowintro.htm

7 D. M. Wegner, R. M. Wenzlaff and M. Kozak (2004). 'Dream rebound: The return of suppressed thoughts in dreams'. *Psychological Science*, 15, pages 232–6.

8 H. A. Murray and D. R. Wheeler (1937). 'A note on the possible clairvoyance of dreams'. *Journal of Psychology*, 3, pages 309–13.

9 C. K. Morewedge and M. I. Norton (2009). 'When dreaming is believing: The (motivated) interpretation of dreams'. *Journal of Personality and Social Psychology*, 96, pages 249–64.

10 K. M. T. Hearne (1978). 'Lucid dreams: an electrophysiological and psychological study'. PhD thesis, University of Hull.

11 The information in this section is based on Stephen LaBerge's 'Mnemonic Induction of Lucid Dreams'.

12 A. Revonsuo (2000). 'The reinterpretation of dreams: An evolutionary hypothesis of the function of dreaming'. *Behavioral and Brain Sciences,* 23, pages 793–1121.

13 F. Crick and G. Mitchison (1983). 'The function of dream sleep'. *Nature,* 304, pages 111–14.

14 P. Mcintyre (2006). 'Paul McCartney and the creation of 'Yesterday': the systems model in operation'. *Popular Music,* 25, pages 201–19.

15 J. A. Hobson and R.W. McCarley (1977). 'The brain as a dream-state generator: An activation-synthesis hypothesis of the dream process'. *American Journal of Psychiatry,* 134, pages 1335–48.

16 M. Solms and O. H. Turnbull (2007). 'To sleep, perchance to REM? The rediscovered role of emotion and meaning in dreams'. In *Tall Tales About the Mind and Brain* (ed. Sergio Della Sala, 2007,) pages 478–500. Oxford University Press, US.

Conclusion

1 J. M. Wood, M. T. Nezworski, S. O. Lilienfeld, H. N. Garb (2002). *What's Wrong With the Rorschach? Science Confronts the Controversial Inkblot Test.* John Wiley & Sons, New York.

2 R. Wiseman and C. Watt (2006). 'Belief in psychic ability and the misattribution hypothesis: A qualitative review'. *British Journal of Psychology,* 97, pages 323–38.
S. J. Blackmore and R. Moore (1994). 'Seeing things: Visual recognition and belief in the paranormal'. *European Journal of Parapsychology,* 10, pages 91–103.
P. Brugger, M. Regard, T. Landis, D. Krebs and J. Niederberger (1994). 'Coincidences: Who can say how "meaningful" they are?' In *Research in parapsychology* (ed. E. W. Cook and D. Delanoy, 1991), pages 94–8. Scarecrow, Metuchen, NJ.

P. Brugger and R. Graves (1998). 'Seeing connections: associative processing as a function of magical belief'. *Journal of the International Neuropsychological Society*, 4, pages 6–7.
R. Wiseman and M. D. Smith (2002). 'Assessing the role of cognitive and motivational biases in belief in the paranormal'. *Journal of the Society for Psychical Research*, 66, pages 178–86.
3 J. Jay (2010). 'Martin Gardner: An Interview'. *Magic Magazine*, 19(11), pages 58–61.
For further information about this aspect of Gardner's thinking, see:
M. Gardner (1983). *The Whys of a Philosophical Scrivener*. Quill, New York.

The Instant Superhero Kit

1 B. R. Forer (1949). 'The fallacy of personal validation: A classroom demonstration of gullibility'. *Journal of Abnormal Psychology*, 44, pages 118–21.
2 G. B. Caputo (2010). 'Strange-face-in-the-mirror illusion', *Perception*, 39(7), pages 1007–08.

Acknowledgements

First and foremost, I wish to thank the University of Hertford-shire for supporting my work over the years. I would like to thank Sue Blackmore, James Randi, Jim Houran, Chris French, Max Maven, the mysterious Mr D, Peter Lamont, and David Britland for their invaluable contributions to this book. Also, special thanks to Emma Greening and Clive Jefferies for reading earlier drafts of the manuscript. This book would not have been possible without the guidance and expertise of my agent Patrick Walsh and editor Jon Butler. Special thanks also to my wonderful colleague, collaborator and partner, Caroline Watt.